GENETICS - RESEARCH AND ISSUES

DNA METHYLATION

PATTERNS, FUNCTIONS AND ROLES IN DISEASE

GENETICS - RESEARCH AND ISSUES

Additional books in this series can be found on Nova's website under the Series tab.

Additional e-books in this series can be found on Nova's website under the eBook tab.

GENETICS - RESEARCH AND ISSUES

DNA METHYLATION

PATTERNS, FUNCTIONS AND ROLES IN DISEASE

KATHLEEN HOLLAND
EDITOR

New York

Copyright © 2016 by Nova Science Publishers, Inc.

All rights reserved. No part of this book may be reproduced, stored in a retrieval system or transmitted in any form or by any means: electronic, electrostatic, magnetic, tape, mechanical photocopying, recording or otherwise without the written permission of the Publisher.

We have partnered with Copyright Clearance Center to make it easy for you to obtain permissions to reuse content from this publication. Simply navigate to this publication's page on Nova's website and locate the "Get Permission" button below the title description. This button is linked directly to the title's permission page on copyright.com. Alternatively, you can visit copyright.com and search by title, ISBN, or ISSN.

For further questions about using the service on copyright.com, please contact:
Copyright Clearance Center
Phone: +1-(978) 750-8400   Fax: +1-(978) 750-4470   E-mail: info@copyright.com.

NOTICE TO THE READER
The Publisher has taken reasonable care in the preparation of this book, but makes no expressed or implied warranty of any kind and assumes no responsibility for any errors or omissions. No liability is assumed for incidental or consequential damages in connection with or arising out of information contained in this book. The Publisher shall not be liable for any special, consequential, or exemplary damages resulting, in whole or in part, from the readers' use of, or reliance upon, this material. Any parts of this book based on government reports are so indicated and copyright is claimed for those parts to the extent applicable to compilations of such works.

Independent verification should be sought for any data, advice or recommendations contained in this book. In addition, no responsibility is assumed by the publisher for any injury and/or damage to persons or property arising from any methods, products, instructions, ideas or otherwise contained in this publication.

This publication is designed to provide accurate and authoritative information with regard to the subject matter covered herein. It is sold with the clear understanding that the Publisher is not engaged in rendering legal or any other professional services. If legal or any other expert assistance is required, the services of a competent person should be sought. FROM A DECLARATION OF PARTICIPANTS JOINTLY ADOPTED BY A COMMITTEE OF THE AMERICAN BAR ASSOCIATION AND A COMMITTEE OF PUBLISHERS.

Additional color graphics may be available in the e-book version of this book.

Library of Congress Cataloging-in-Publication Data

ISBN: 978-1-53610-244-4

Published by Nova Science Publishers, Inc. † New York

Contents

Preface vii

Chapter 1 Disruption of DNA Methylation Patterns Caused by Exposure to Environmental Factors during the Developmental Period 1
Ken Tachibana and Ken Takeda

Chapter 2 Cytosine Modifications Affect Biological Significance by a Multifactorial Network 29
Selcen Çelik-Uzuner

Chapter 3 Targeting Aberrant IGF2 Epigenetics as a Novel Anti-Tumor Approach 91
Ji-Fan Hu and Andrew R. Hoffman

Bibliography 111

Related Nova Publications 191

Index 197

PREFACE

This book provides new research on the patterns, functions and roles in disease of DNA methylation. Chapter One studies the disruption of DNA methylation patterns caused by exposure to environmental factors during the developmental period. Chapter Two presents transcriptional regulation by cytosine modifications through a broad concept of molecular sciences in physics, chemistry and biology, and suggests a complex network for cytosine modifications-mediated transcription depending on multiple factors. Chapter Three reviews the loss of Insulin-like growth factor 2 (IGF2) imprinting in human tumors, focusing especially on the mechanisms underlying this abnormality in epigenetic control; and summarizes recent progress on targeting this tumor specific imprinting abnormality as a novel anti-cancer approach.

Chapter 1 – DNA methylation is a critical mechanism of epigenetic gene regulation. The methylation status of CpG islands, which are GC-rich DNA regions that possess relatively high densities of CpG dinucleotide, is closely associated with gene transcription activity. Knockout studies of DNA methyltransferases, which are responsible for the methylation of cytosine residue at CpG sites, have shown that DNA methylation is essential for complete embryonic development. In the developmental period, the DNA methylation pattern derived from germ cells disappears when the fertilized egg develops into a blastocyst. The *de novo* DNA methylation pattern is then reestablished at around the stage of implantation. The global DNA methylation level also changes in the early postnatal stages. These DNA methylation processes that occur during development are associated with long-lasting phenotypic changes, including genomic imprinting, cell differentiation, and X-chromosome inactivation. This dynamic regulation of the DNA methylation status during the developmental period allows for normal development. Disruption of the DNA methylation pattern by exposure to chemicals/metals during the developmental period is suspected to affect the development of offspring. In fact, several reports have shown that prenatal exposure to chemicals/metals causes developmental defects in the offspring through disruption of the DNA methylation status. Recently, the authors found that exposure to prenatal diesel exhaust, which contains particulate matter, disrupts the DNA methylation status of the brain of mouse offspring. In addition, it has been revealed that various environmental factors have the ability to alter DNA methylation status in the developmental stages. Further investigations will be required to elucidate the mechanism underlying the disruption of DNA methylation caused by prenatal exposure to chemicals and particulate matter and to avoid the detrimental health effects associated with aberrant DNA methylation.

Chapter 2 – Modifications at cytosine base are of great interests since they have been shown to alter during lifespan of many organisms and to involve in transcriptional regulation. The first discovered one, 5'-methylcytosine (5meC), can be converted to other modifications, 5'-hydroxymethylcytosine (5hmC), 5'-formylcytosine (5fC) and 5'-carboxycytosine (5caC) by enzymatic reactions, and the reversible conversions are also possible. This represents a model for epigenetic dynamism which is such represented by the hypothesis for active demethylation mediated by DNA repair. Some abnormalities associated with the changes in cytosine modifications suggest their importance for biological processes. Although there is a limited understanding of the biological importance of 5fC and 5caC, 5meC has been shown to associate with gene silencing, but 5hmC is commonly thought to involve in gene activation. Though this is not that simple since the effects of cytosine modifications can depend on such their genomic location, being within cytosine-guanine repeats or not, the sequence of DNA part they are included, in association with three dimensional features of DNA structure indicating how cytosine modifications affect transcriptional outcome is hard to understand. This chapter presents transcriptional regulation by cytosine modifications through a broad concept of molecular sciences in physics, chemistry and biology, and suggests a complex network for cytosine modifications-mediated transcription depending on multiple factors.

Chapter 3 – Insulin-like growth factor 2 (*IGF2*) encodes a potent fetal mitogen that regulates cell proliferation, growth, differentiation and survival. In normal tissues, the gene is maternally imprinted, and its expression is epigenetically regulated by the coordination of differential allelic DNA methylation in the imprinting control region, CTCF/SUZ12-mediated intrachromosomal looping, and allelic chromatin histone modifications in the gene's promoters. Loss of *IGF2* imprinting has been observed in a variety of growth disorders and malignant tumors. This epigenetic mutation may provide investigators with novel biomarkers for early cancer detection, prediction, and prognosis. Targeting this tumor-specific epigenetic abnormality may represent a potential approach for the development of novel therapeutic anti-cancer strategies.

In: DNA Methylation
Editor: Kathleen Holland

ISBN: 978-1-53610-244-4
© 2016 Nova Science Publishers, Inc.

Chapter 1

DISRUPTION OF DNA METHYLATION PATTERNS CAUSED BY EXPOSURE TO ENVIRONMENTAL FACTORS DURING THE DEVELOPMENTAL PERIOD

Ken Tachibana[1,2] and Ken Takeda[2]
[1]Department of Health Biosciences,
Nihon Pharmaceutical University, Japan
[2]The Center for Environmental Health Science for the Next Generation,
Research Institute for Science and Technology,
Tokyo University of Science, Japan

ABSTRACT

DNA methylation is a critical mechanism of epigenetic gene regulation. The methylation status of CpG islands, which are GC-rich DNA regions that possess relatively high densities of CpG dinucleotide, is closely associated with gene transcription activity. Knockout studies of DNA methyltransferases, which are responsible for the methylation of cytosine residue at CpG sites, have shown that DNA methylation is essential for complete embryonic development.

In the developmental period, the DNA methylation pattern derived from germ cells disappears when the fertilized egg develops into a blastocyst. The *de novo* DNA methylation pattern is then reestablished at around the stage of implantation. The global DNA methylation level also changes in the early postnatal stages. These DNA methylation processes that occur during development are associated with long-lasting phenotypic changes, including genomic imprinting, cell differentiation, and X-chromosome inactivation. This dynamic regulation of the DNA methylation status during the developmental period allows for normal development.

Disruption of the DNA methylation pattern by exposure to chemicals/metals during the developmental period is suspected to affect the development of offspring. In fact, several reports have shown that prenatal exposure to chemicals/metals causes developmental defects in the offspring through disruption of the DNA methylation status. Recently, the authors found that exposure to prenatal diesel exhaust, which contains particulate matter, disrupts the DNA methylation status of the brain of mouse offspring. In addition, it has been revealed that various environmental factors have the ability to alter DNA methylation status in the developmental stages. Further investigations will be required to elucidate the mechanism underlying the disruption of DNA methylation caused by prenatal exposure to chemicals and particulate matter and to avoid the detrimental health effects associated with aberrant DNA methylation.

Keywords: DNA methylation, endocrine disrupting chemicals, metals, maternal diet, cigarette smoke, particulate matter, developmental period

INTRODUCTION

Human epidemiologic and animal studies indicate that nutrition and environmental stimuli during prenatal and postnatal mammalian development influence developmental pathways and thereby induce permanent changes in metabolism and susceptibility to chronic disease (McMillen and Robinson 2005). Nutritional intake during pregnancy was the first factor identified that affects fetal development (Barker et al. 1993). Since then, many studies have shown that environmental factors are closely associated with reproductive and child health. These phenomena led to the proposal of the "developmental origins of health and diseases (DOHaD)" hypothesis (Barker 1995). Epigenetic mechanisms likely play an important role in this hypothesis (Waterland, R. A. and Michels 2007).

DNA methylation is a critical mechanism of epigenetic gene regulation (Deaton and Bird 2011). In mammals, methylation almost exclusively occurs on the cytosine residue of CpG dinucleotide. CpG islands are GC-rich DNA regions that possess relatively high densities of CpG dinucleotide. They are found in many genes, positioned mainly at the 5' ends. Their methylation status is closely associated with gene transcription activity. DNA methylation results in transcriptional silencing, either by interfering with transcription factor binding or by inducing a heterochromatin structure. DNA methyltransferases DNMT1, DNMT3a, and DNMT3b are responsible for the methylation of cytosine residue in CpG sites (Bergman and Cedar 2013). DNMT3a and DNMT3b play crucial roles in *de novo* cytosine methylation, whereas DNMT1 works by maintaining the DNA methylation pattern in the newly synthesized DNA during cell division. Knockout studies of DNA methyltransferases have shown that DNA methylation is essential for complete embryonic development (Li et al. 1992, Okano et al. 1999).

In the developmental period, the DNA methylation pattern derived from germ cells disappears when the fertilized egg develops into a blastocyst. The *de novo* methylation pattern is then reestablished at around the stage of implantation (Kafri et al. 1992). The global DNA methylation level also changes in the early postnatal stages (Tawa et al. 1990). This dynamic regulation of the DNA methylation status during the developmental period would provide a mechanism for the removal of errors in gene methylation patterns derived from germ lines, thus allowing normal development. These DNA methylation processes that occur during development are associated with long-lasting phenotypic changes, including genomic imprinting, cell differentiation, and X-chromosome inactivation (Roth et al. 2009). In addition, tissue-specific DNA methylation patterns are constructed during this developmental period (Liang et al. 2011), and this differential DNA methylation would be crucial for each organ to function properly. Aberrant DNA methylation is associated with various diseases and disorders. In fact, previous studies indicated that dysregulation of DNA methylation contributes to immunodeficiency, centromeric region instability, and facial anomalies syndrome (Sutcliffe et al. 1992, Amir et al. 1999, Tucker 2001).

The developmental embryo/fetus could be highly vulnerable to environmental stimuli, even those having no toxic effect on the adult. The majority of environmental factors do not appear to change the nucleotide sequence of DNA, that is, these environmental factors do not induce genetic mutations (McCarrey 2012). However, environmental factors can affect the epigenetic modifications, which include DNA methylation and development. In fact, previous reports suggested that prenatal exposure to environmental factors, such as chemicals, metals, or particulate matter, disrupts the DNA methylation pattern and subsequently induces various abnormal phenotypes in the offspring.

This chapter introduces the research into the effects of various environmental factors on DNA methylation and discusses possible mechanisms by which exposure to environmental factors induces the disruption of DNA methylation.

ENDOCRINE-DISRUPTING CHEMICALS

Although the definition of endocrine-disrupting chemicals (EDCs) is complicated and controversial (Foster and Agzarian 2008), they are generally defined as environmental chemicals that can interfere with any aspect of hormone function (Fowler et al. 2012, Zoeller et al. 2012, Bergman et al. 2013). The endocrine disruptors are often associated with specific classes of hormones. For example, bisphenol A (BPA), diethylstilbestrol, and genistein have the ability to disrupt estrogenic signaling (Kelce et al. 1994). Organic compounds such as hydrocarbons or 2,3,7,8-tetrachlorodibenzo-p-dioxin, which bind to aryl hydrocarbon receptors, also disrupt this signaling (Skinner 2014). To date, a number of EDCs have been shown to disrupt DNA methylation status (Table 1).

The anti-androgenic compound vinclozolin, which is one of the most commonly used agricultural fungicides, was the first EDC found to disrupt the DNA methylation (Anway et al. 2005). Vinclozolin that was transiently exposed to pregnant rats (F0 generation) decreased the spermatogenic capacity (cell number and viability) in the F1 generation. Anway et al. (2005) showed that the effects on reproduction correlated with altered DNA methylation patterns of lysophospholipase and cytokine-inducible SH2 protein in the male germ line. Altered DNA methylation patterns appeared to be transgenerationally transmitted to epididymal sperm of the F3 generation.

BPA and phthalates are important plasticizers. Previous study showed that prenatal BPA exposure affects sexual differentiation and behavior. Kundakovic et al. (2013) indicated that prenatal exposure to environmentally relevant doses of BPA induce sex-specific, dose-dependent, and brain region-specific changes in the expression of estrogen receptors (ERs) and estrogen-related receptor-γ in mouse offspring. Prenatal BPA also altered the mRNA expression level of DNMT 1 and DNMT3a in the brain. Importantly, these alterations in ERα and DNMT expression were associated with DNA methylation changes in the ERα gene. This study also demonstrated that prenatal BPA exposure induces the lasting disruption of DNA methylation in mouse offspring. Prenatal BPA exposure also disrupts genomic imprinting in the mouse (Susiarjo et al. 2013). Imprinted genes are regulated by differential DNA methylation, and disrupted imprinting causes adverse effects on placental, fetal, and postnatal development

(Charalambous et al. 2007). Prenatal BPA exposure during late stages of oocyte development and early stages of embryonic development significantly altered the DNA methylation levels of differentially methylated regions (DMRs) of *Snrpn* and *Igf2* (Susiarjo et al. 2013).

Table 1. Altered DNA methylation caused by prenatal exposure to endocrine disruptors

Exposure*	Model	Gene	DNA methylation	Tissue/Cells	Reference
Vinclozolin	Rat	Lysophospholipase Cish	Decreased methylation	Testis	Anway et al. 2005
n-Butylparaben	Rat	ERα	Decreased methylation	Testis	Zhang et al. 2016
BPA	Rat	Fkbp5	Increased methylation	Hypothalamus Hippocampus	Kitraki et al. 2015
BPA	Mouse	ERα	Increased methylation Decreased methylation	Cortex (Male) Hypothalamus (Female)	Kundakovic et al. 2013
BPA	Mouse	Igf2 Snrpn	Increased methylation Decreased methylation	Placenta Embryo (E9.5)	Susiarjo et al. 2013
DEHP	Rat	-	Global hydroxymethylation	Testis	Abdel-Maksoud et al. 2015
DEHP	Rat	-	Globally altered methylation	Adrenal gland	Martinez-Arguelles and Papadopoulos. 2015
DEHP	Mouse	Igf2r Peg3	Decreased methylation	Primordial germ cell	Li et al. 2014
TBTCl	Mouse	-	Global hypomethylation	Oocyte	Huang et al. 2015
BDE-47	Rat	Mt-co2 Bdnf Nr3c1	Decreased methylation	Frontal lobes	Byun et al. 2015
TCDD	Mouse	Igr2r	Increased methylation	Muscle Liver	Somm et al. 2013
DES	Mouse	Casq2	Increased methylation	Heart	Haddad et al. 2013

*Abbreviations: BPA, bisphenol A; DEHP, di(2-ethylhexyl) phthalate; TBTCl, tributyltin chloride; BDE-47, 2,2',4,4'-tetrabromodiphenyl ether; TCDD, 2,3,7,8-tetrachlorodibenzo-*p*-dioxin; DES, diethylstilbestrol

In addition, several studies showed that prenatal phtarate exposure alters DNA methylation patterns. Prenatal exposure to di(2-ethylhexyl) phthalate (DEHP) from GD12.5 to 19 has been shown to increase global gene methylation in the testis of the mouse (Wu et al. 2010). Another report indicated that prenatal exposure to DEHP significantly reduced DNA methylation in *Igf2r* and *Peg3* DMRs in primordial germ cells from male and female fetal mice (Li et al. 2014). The disrupted DNA methylation of imprinted genes in the F1 mouse oocyte was heritable to the F2 offspring (Li et al. 2014). Interestingly, Martinez-Arguelles and Papadopoulos (2015) identified hot spots of DNA methylation changes primarily within CpG islands followed by shelf regions (2-4 kb away from a CpG island), which are known to control regions for gene expression. These findings indicated that prenatal exposure to EDCs causes disruption of DNA methylation patterns in the offspring; however, further investigation is required to clarify their role in the adverse effects on offspring.

HEAVY METALS

A number of reports indicated that exposure to heavy metals induces alteration of DNA methylation patterns (Table 2). Cadmium, which is considered a partial transplacental toxic metal, is ubiquitous in the environment, with primary exposure occurring via dietary intake, cigarette smoke, and industrial emissions (Olsson et al. 2002, Järup and Åkesson 2009). Pregnant women appear to accumulate more cadmium than nonpregnant women (Nishijo et al. 2004). Exposed cadmium is accumulated in the placenta and has been considered as a potential cause of adverse effects in offspring (Ji et al. 2011, Kippler et al. 2012, Sakamoto et al. 2013). Castillo et al. (2012) investigated the effects of cadmium exposure through drinking water at 50 ppm cadmium administered *ad libitum* to pregnant rats on DNA methylation of the glucocorticoid receptor in the liver of the offspring. Interestingly, male and female offspring showed opposite effects in this study. Male offspring showed hypermethylation of the glucocorticoid receptor accompanied by lower mRNA expression, whereas the female offspring showed hypomethylation and overexpression of the glucocorticoid receptor gene. Coinciding with these results, sex-specific alterations in DNA methylation are observed in human newborns exposed to cadmium in early life (Kippler et al. 2013).

Table 2. Altered DNA methylation caused by prenatal exposure to metals

Exposure	Model	Gene	DNA methylation	Tissue/Cells	Reference
Cadmium	Rat	Glucocorticoid receptor	Increased methylation Decreased methylation	Liver (Male) Liver (Female)	Castillo et al. 2012
Cadmium	Human	-	Global hypermethylation Global hypomethylation	Cord blood (Male) Cord blood (Female)	Kippler et al. 2013
Methyl mercury	Mouse	BDNF	Increased hypermethylation	Hippocampus	Onishchenko et al. 2008
Methyl mercury	Rat	Cdkn2a	Decreased methylation	Liver	Desaulniers et al. 2009
Lead	Mouse	About 150 genes	Increased methylation	Brain	Dosunmu et al. 2012
Lead	Rat	-	Altered methylation (related to altered DNMT1 and DNMT3a)	Hippocampus	Schneider et al. 2013
Lead	Human	NDRG4, APOA5 NINJ2, TRPV2, DOK3	Increased methylation Decreased methylation	Neonatal blood	Sen et al. 2015
Arsenic	Human	-	Global hypermethylation	Cord blood (CD8[+] T lymphocytes)	Koestler et al. 2013
Arsenic	Human	71 genes (ZNF710, TNFRSF10B, etc.)	Increased methylation	Cord blood	Kile et al. 2014
Zinc	Chick	A20	Decreased methylation	Gut mucosa	Li et al. 2015

Methylmercury (MeHg) is a neurotoxic organic form of the environmental toxicant mercury. Humans can be exposed to MeHg by consuming diets containing contaminated seafood. Because MeHg readily crosses the placenta, the developing fetal nervous system is especially affected (Myers and Davidson 1998). Prenatal exposure to high-dose MeHg causes mental retardation and cerebral palsy. Prenatal MeHg exposure in mice has been shown to lead to DNA hypermethylation, increased histone tri-methylation, and decreased histone acetylation within the promotor of the brain-derived neurotrophic factor (*BDNF*) gene in the hippocampus of offspring and is associated with depressive-like behaviors (Onishchenko et al. 2008). Furthermore, Desaulniers et al. (2009) reported that the offspring of rats prenatally exposed to MeHg (from gestation day 1 to postnatal day 21 at 2 mg/kg) showed a significant decrease in the mRNA levels of *Dnmt1* and *Dnmt3b*. Cyclin-dependent kinase inhibitor 2A (*cdkn2a*), which encodes the tumor suppressor p16^{INK4a} protein, showed a significant decrease in methylation status in the promoter region. These offspring showed hepatic inflammation, vacuolation, and hypertrophy. Therefore, prenatal and postnatal exposure to MeHg would lead to neurological disorder and liver toxicity through the alteration of DNA methylation.

Lead (Pb) is one of the most widely used environmental toxicants, and its exposure causes various adverse effects in the neuronal and reproductive systems (Kašuba et al. 2010). A study of genome-wide DNA methylation and gene expression in the brain of mice prenatally exposed to Pb revealed a significant association between the increase in DNA methylation and transcriptional repression of genes (Dosunmu et al. 2012). About 150 genes were differentially expressed in the elderly offspring. These genes, which are related to the immune response, metal binding, metabolism, and transcription/transduction coupling, are upregulated in normal aging. Dosunmu et al. argued that prenatal exposure to Pb revealed repression in these genes suggesting that disturbances in the developmental stages of the brain compromise the ability to defend against age-related stressors, thus promoting the neurodegenerative process. Altered DNMT expression would be associated with the effects of prenatal Pb exposure on neuronal development (Schneider et al. 2013). Another study showed that Pb exposure during pregnancy affects the DNA methylation status of neonatal blood cells (Sen et al. 2015). This altered DNA methylation was transgenerationally inherited by the grandchildren.

Inorganic arsenic (As) compounds are known to be environmental toxicants that affect DNA methylation status. People are primarily exposed to As from drinking water, industrial emissions, and dietary sources (Gilbert-Diamond et al. 2011, Green and Marsit 2015). Because As is methylated during its metabolism, As metabolism may deplete intracellular methyl group storage, which may result in hypomethylation. However, in fact, both hypo- and hypermethylation of different genes can be caused by exposure to As (Zhong and Mass 2001). Several investigations of the relationship between prenatal As exposure and DNA methylation in both umbilical cord blood and maternal leukocytes were performed (Koestler et al. 2013, Kile et al. 2014). Kile et al. (2014) examined the relationship between As in maternal drinking water and DNA methylation in cord blood adjusting for leukocyte-tagged DMRs. This study indicated that the increase in the population of CD8$^+$ T cells was accompanied by differential DNA methylation throughout the genome. Because As-associated alteration of

DNA methylation was enriched in CpG islands, these alterations may affect the expression of genes.

Although a number of publications have reported the adverse effects of prenatal exposure to heavy metals, little is known about the involvement of DNA methylation in these effects. Further examinations are required to clarify the molecular basis underlying the adverse effects caused by prenatal exposure to heavy metals.

PARTICULATE MATTER

Epidemiological and experimental studies have shown that exposure to fine ambient particulate matter is related to respiratory and cardiovascular disorders (Pope et al. 2004, Ostro et al. 2006, Liu et al. 2008). Because diesel exhaust (DE) is one of the main types of air pollution and is a major source of fine ambient particulate matter in urban environments (Donaldson et al. 2005), models of DE exposure have been used to investigate the health effects of ambient particulate matter.

Several studies have indicated that exposure to DE may affect the central nervous system. For instance, railroad workers exposed to DE have shown neurobehavioral impairment (Kilburn 2000), and human volunteers exposed to DE have shown altered electrical signals in the frontal cortex (Crüts et al. 2008). Suzuki et al. (2010) reported that prenatal exposure to DE affects the brain of offspring with regard to neurotransmitter levels and spontaneous locomotor activity. Other studies have shown that prenatal exposure to DE induces neuroinflammation and affects behavior in mouse offspring (Bolton et al. 2012, Thirtamara Rajamani et al. 2013). Peters et al. (2013) indicated that prenatal exposure to DE increases the risk of childhood brain tumors. Although these reports suggest that exposure to DE affects the brain of offspring in the developmental period, the molecular events involved in these health effects has largely remained unclear.

Recently, the authors analyzed the effects of prenatal exposure to DE on DNA methylation status in the brain of offspring mice (Tachibana et al. 2015). The methylation status of the promoter DNA region throughout the entire genome was analyzed. The methylated promoter DNA regions specific to the control group (which were defined as those DNA methylation levels in the DE-exposed group that were decreased in comparison to the control group) and specific to the DE-exposed group (which were defined as those DNA methylation levels in the DE-exposed group that were increased in comparison to the control group) were analyzed. Altered DNA methylation on the promoter regions was detected in all chromosomes in 1-day-old (1d) male, 1d female, 21d male, and 21d female offspring (Figures 1, 2). These results indicated that prenatal exposure to DE disrupted the genome-wide DNA methylation state in the brain of offspring mice throughout the 1-21-day postnatal period. The differentially methylated genes were then categorized bioinformatically using Gene Ontology (GO) to identify the molecular events associated with altered DNA methylation induced by prenatal exposure to DE. To obtain information about the biological function affected by altered DNA methylation at each time point, GO terms that were common between male and female offspring were extracted. The GO terms related to neuronal differentiation ("positive regulation of neuron differentiation") and neurogenesis ("positive regulation of neurogenesis" and "neurogenesis") were found in both the 1d and 21d offspring, respectively. These results suggest that disrupted DNA methylation in the infertile mouse brain is involved in neural dysfunctions induced by prenatal exposure to DE. A human crossover study showed that DE inhalation is associated with changes in DNA methylation of mononuclear cells in asthmatics (Jiang et al. 2014). This study indicated that DE has an ability to alter the DNA methylation pattern not only in mice but also humans.

As the production and use of nanomaterials (NMs) continues to expand, the potential risk of toxicity to humans and the environment caused by NMs increases (Johnston et al. 2013). Several studies indicate that exposure to NMs induces not only genomic mutations but also epimutations, which include altered DNA methylation. Some human population studies suggest that exposure to NMs can affect the DNA methylation pattern. Baccarelli et al. (2009) reported that exposure to black carbon, which is produced by traffic, is associated with a decrease in the methylation of long interspread nuclear element-1 (LINE-1) in blood cells of elderly participants. Similarly, Madrigano et al. (2011) indicated that decreased methylation of LINE-1 is associated with exposure to black carbon. Although these reports show the ability of NMs to affect the DNA methylation pattern in adults, it remains unclear whether prenatal exposure to NMs affects the construction of precise DNA methylation patterns in the developmental stage. A recent report revealed that prenatal exposure to silver nanoparticles induces the aberrant expression of imprinted genes, such as *Ascl2*, *Snrpn*, *Peg3*, *Kcnq1ot1*, *Zac1*, *H19*, *Igf2r*, and *Igf2*, in placenta (Zhang et al. 2015). Zhang et al. (2015) also showed that downregulation of the methylation of *Zac1* DMR may be associated with the upregulation of *Zac1* gene expression. Methylation of *Igf2* DMR was slightly increased in the silver nanoparticle-treated group. The *Zac1* gene changes the expression of imprinted genes including *H19*, *Igf2*, *Cdkn1c*, and *Dlk1*, and also directly regulates the *Igf2/H19* locus through binding to a shared enhancer (Varrault et al. 2006). Because the *Zac1* gene is critically involved in the control of embryonic growth, aberrant expression of the *Zac1* gene, which is associated with altered DNA methylation in the placenta, could potentially explain the abnormal fetal growth.

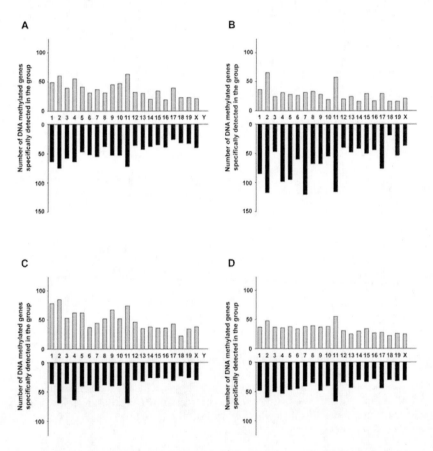

Figure 1. Effects of prenatal diesel exhaust (DE) exposure on the DNA methylation state of gene promoters with CpG islands. The number of DNA methylated genes specifically detected in the control and DE groups in 1-day-old (1d) male (A), 1d female (B), 21d male (C), and 21d female (D) offspring are shown. The x-axis shows chromosome numbers. Black bars indicate the genes specifically methylated in the control (meaning that DNA methylation was decreased by DE exposure). Gray bars indicate the genes specifically methylated in the DE groups (meaning that DNA methylation was increased by DE exposure).

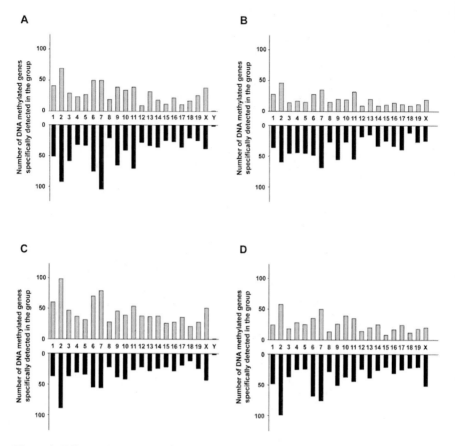

Figure 2. Effects of prenatal diesel exhaust (DE) exposure on the DNA methylation state of gene promoters without CpG islands. The number of DNA methylated genes specifically detected in the control and DE groups in 1-day-old (1d) male (A), 1d female (B), 21d male (C), and 21d female (D) offspring are shown. The x-axis shows chromosome numbers. Black bars indicate the genes specifically methylated in the control (meaning that DNA methylation was decreased by DE exposure). Gray bars indicate the genes specifically methylated in the DE groups (meaning that DNA methylation was increased by DE exposure).

FOOD AND ALCOHOL INTAKE

A number of studies have revealed that the maternal diet alters the DNA methylation pattern of offspring. Ruchat et al. (2013) showed that gestational diabetes mellitus during pregnancy induces altered DNA methylation in placental tissue and cord blood. These differentially methylated genes were likely involved in the metabolic diseases pathway and were also associated with newborn weight. Another study focused on the association of maternal obesity with methylation of the leptin gene, which is important for metabolism (Lesseur et al. 2013). The study indicated that both cord blood and placental tissue had significantly higher methylation of the leptin gene in the case of infants born small for their gestational age. Dunn and Bale (2009) found that a maternal high-fat diet promotes body length increase and increased insulin sensitivity in second-generation mice. DNA methylation of the growth hormone secretagogue receptor (*GHSR*) gene, which is associated with the growth hormone axis, was significantly decreased in these mice. In addition, a cohort study suggested that higher methylation of retinoid X receptor-α was associated with lower maternal carbohydrate intake in early pregnancy and was linked with higher neonatal adiposity (Godfrey et al. 2011). These data provide supportive evidence that DNA methylation is involved in fetal metabolic programming and potentially predisposes the fetus to metabolic diseases.

Dietary intake of folates has important effects on brain development during pregnancy. Decreased folate concentrations in serum and red cells are responsible for causing clinical folate deficiency (Zeisel 2009). Pregnant women with lower erythrocyte folate concentrations are more likely to have a baby with neural tube birth defects (Smithells et al. 1976). Low dietary intake of folate not only depletes folate metabolites but also decreases S-adenosylmethionine concentrations (Shivapurkar and Poirier 1983), with resulting hypomethylation of DNA (Tsujiuchi et al. 1999). A methyl donor-deficient diet during the developmental period can affect fear and anxiety in adulthood in mice (Ishii et al. 2014). These neuronal disorders appear to be associated with aberrant expression of the *Dnmt3a* and *Dnmt3b* genes, which are critical for *de novo* DNA methylation. In contrast, Hollingsworth et al. (2008) showed that *in utero* supplementation of dietary methyl donors, which contained folic acid, caused the reduced expression of Runt-related transcription factor (*Runx3*), a gene known to negatively regulate allergic airway disease, by excessive DNA methylation. Another study showed that supplementation of folate after 12 weeks of gestation affects the DNA methylation status of imprinted genes in cord blood (Haggarty et al. 2013). This study indicated that folate supplementation is associated with a higher level of methylation in *IGF2* and reduction in *PEG3* and LINE-1. Aberrant methylation of imprinted genes has also been reported (Liu et al. 2012, Hoyo et al. 2014). The mechanism of altered DNA methylation of imprinting genes in the presence or absence of folate may not be due to methyl group bioavailability (Green and Marsit 2015). Amarasekera et al. (2014) reported that neonate CD4$^+$ T cells and antigen-presenting cells exposed to either a high- or low-folate diet have differential methylation in the upstream region of the zinc finger protein-57 (*ZFP57*) gene. Because the *ZFP57* gene is crucial for the establishment and maintenance of imprinting-associated DNA methylation (Quenneville et al. 2011, Zuo et al. 2012), differential methylation of the imprinting gene caused by maternal folate intake might be associated with altered regulation of *ZFP57*.

It is widely known that maternal alcohol intake induces a wide range of developmental abnormalities in fetuses. These disorders are collectively referred to as fetal alcohol spectrum disorders (FASD) (Astley 2004). FASD is known to associate with three features: prenatal/postnatal growth retardation, distinctive facial features, and brain damage (Hoyme et al. 2005). Recent research has demonstrated the potential involvement of aberrant DNA methylation in FASD. Genome-wide analysis of gene expression in the mouse hippocampus revealed altered 23 mRNA and 3 microRNA expression in offspring prenatally exposed to ethanol at postnatal day 28 (Marjonen et al. 2015). In addition, altered DNA methylation was observed in the CpG islands of these differentially expressed genes. Another study showed that prenatal ethanol exposure decreased the expression level of proopiomelanocortin (POMC) but increased the level of DNA methylation in the hypothalamus in which the POMC-positive neuronal cell bodies are located in rat (Gangisetty et al. 2014). A significant increase in methyl CpG binding protein (MeCP2) expression and increased MeCP2 protein binding on the POMC promoter were also observed at this time. These alterations might contribute to dysfunction of the hypothalamic-pituitary-adrenal axis. These reports suggest that dysregulated DNA methylation and subsequent altered gene expression are critically involved in FASD.

CIGARETTE SMOKE

Prenatal exposure to cigarette smoke has been associated with a number of adverse effects in the exposed offspring (Oken et al. 2008, Power et al. 2010, Syme et al. 2010). Cigarette smoke is a complex mixture of more than 4,700 chemical compounds (MacNee 2005) such as carcinogens, nicotine, and carbon monoxide that have been shown to modify DNA methylation in differentiating and dividing cells (Mortusewicz et al. 2005, Cuozzo et al. 2007, Satta et al. 2008). Markunas et al. (2014) performed an epigenome-wide association study using the whole blood of infants whose mothers smoked during the first trimester. In this study, 185 CpGs with altered methylation were identified in the infants of smokers. These CpG sites correspond to genes such as *FRMD4A*, *ATP9A*, *GALNT2*, and *MEG3*, which are implicated in the processes related to nicotine dependence, smoking cessation, and placental and embryonic development. A cohort study found that 5 CpGs were associated with prenatal exposure to cigarette smoke at birth and during childhood, and 4 of the 5 CpGs were associated with prenatal exposure to cigarette smoke during adolescence (Lee et al. 2015). These studies clearly showed that the DNA methylation pattern is altered by exposure to cigarette smoke during the developmental period. Furthermore, this altered DNA methylation in the infant may persist at least until adolescence.

POSSIBLE MECHANISM OF DYSREGULATION OF DNA METHYLATION

The molecular mechanisms underlying the effect of exposure to prenatal environmental factors on the DNA methylation patterns remain unknown. Several reports indicate a relationship between reactive oxygen species (ROS) and DNA methylation. Metals, nanoparticles (including DE-derived particles), and cigarette smoke are well known to generate ROS in tissue and cells (Nel et al. 2006, Beyersmann and Hartwig 2008, Kelsen et al. 2008, Li et al. 2010). Furthermore, several EDCs also produce ROS in human or rat cells (Okai et al. 2004, Xin et al. 2014). Oxidation of guanine in CpG dinucleotide reduced the binding of the methyl-CpG binding domain of MeCP2 to that site (Valinluck et al. 2004). Kimura and Shiota (2003) showed that genome-wide and/or -specific local DNA methylation may be maintained by the Dnmt1-MeCP2 complexes bound to hemimethylated DNA during replication. Taken together, the obstruction of MeCP2 binding to DNA induced by oxidative damage may possibly alter DNA methylation.

In addition, the biological systems that determine the DNA regions that are methylated are another possible target of exposure to environmental factors. Previous reports indicated several factors essential for the establishment and maintenance of the methylation imprint, including Zfp57 and PGC7/Stella (Nakamura et al. 2007, Li et al. 2008). Shen et al. (2013) showed that a dynamic methylation-demethylation cycle occurs at a large number of genomic loci. This hydroxymethylation, which was especially found in embryonic stem cells and brain (Kriaucionis and Heintz 2009, Ruzov et al. 2011), may play important roles in DNA demethylation in this cycle (Tahiliani et al. 2009, Guo et al. 2011).

These molecules and pathways would also be candidate targets of prenatal exposure to environmental factors. Recently, a portion of the piRNA, small RNA exclusively expressed in the germ line, was linked to *de novo* DNA methylation (Olovnikov et al. 2012). In contrast, Weaver et al. (2005) showed that maternal stress alters the epigenotype in rodent offspring. Analysis of whether environmental factors disrupt the molecules/pathways indicated above would help to solve the problem.

CONCLUSION

There is growing evidence that prenatal exposure to environmental factors can promote dysregulation of DNA methylation and subsequent alteration of gene expression. This evidence indicates that the early-life social environment could be critical for the construction of the DNA methylation pattern and may be associated with a long-term effect on health. Furthermore, it was reported that DNA hypomethylation induces genomic instability (Chen et al. 1998). Therefore, DNA hypomethylation, which is induced by conditions such as low folate status (Jacob et al. 1998, Yi et al. 2000), may lead to an increase in the mutation rate. To avoid the adverse effects caused by exposure to environmental factors, it is important to clarify the molecular mechanisms of these effects.

ACKNOWLEDGMENTS

This work was supported in part by a JSPS KAKENHI for Young Scientist (B) (Grant Number 22710067, 2010-2011), JSPS KAKENHI Grant-in-Aid for Scientific Research (B) (Grant Numbers 21390037, 2009-2011; 24390033, 2012-2014), JSPS KAKENHI Grant-in-Aid for Scientific Research (C) (Grant Numbers 24510085, 2012-2015; 15K00558, 2015-2017); a MEXT-Supported Program for the Strategic Research Foundation at Private Universities (Grant Number S1101015, 2011-2015); and a Grant-in-Aid for the Health and Labour Science Research Grant, Research on the Risk of Chemical Substances (Grant Number 12103301, 2012-2014) from the Ministry of Health, Labour and Welfare of Japan. The funders had no role in either the preparation of or decision to publish the manuscript. We thank the staff and the graduate and undergraduate students at the Takeda laboratory for their help with the experiments.

REFERENCES

Abdel-Maksoud, F. M., Leasor, K. R., Butzen, K., Braden, T. D. and Akingbemi, B. T. (2015). Prenatal Exposures of Male Rats to the Environmental Chemicals Bisphenol A and Di(2-Ethylhexyl) Phthalate Impact the Sexual Differentiation Process. *Endocrinology, 156*, 4672-4683.

Amarasekera, M., Martino, D., Ashley, S., Harb, H., Kesper, D., Strickland, D., Saffery, R. and Prescott, S. L. (2014). Genome-wide DNA methylation profiling identifies a folate-sensitive region of differential methylation upstream of ZFP57-imprinting regulator in humans. *FASEB J, 28*, 4068-4076.

Amir, R. E., Van den Veyver, I. B., Wan, M., Tran, C. Q., Francke, U. and Zoghbi, H. Y. (1999). Rett syndrome is caused by mutations in X-linked MECP2, encoding methyl-CpG-binding protein 2. *Nat Genet, 23*, 185-188.

Anway, M. D., Cupp, A. S., Uzumcu, M. and Skinner, M. K. (2005). Epigenetic transgenerational actions of endocrine disruptors and male fertility. *Science, 308*, 1466-1469.

Astley, S. J. (2004). Fetal alcohol syndrome prevention in Washington State: evidence of success. *Paediatr Perinat Epidemiol, 18*, 344-351.

Baccarelli, A., Wright, R. O., Bollati, V., Tarantini, L., Litonjua, A. A., Suh, H. H., Zanobetti, A., Sparrow, D., Vokonas, P. S. and Schwartz, J. (2009). Rapid DNA methylation changes after exposure to traffic particles. *Am J Respir Crit Care Med, 179*, 572-578.

Barker, D.J. (1995). Fetal origins of coronary heart disease. *BMJ, 311*, 171-174.

Barker, D. J., Gluckman, P. D., Godfrey, K. M., Harding, J. E., Owens, J. A. and Robinson, J. S. (1993). Fetal nutrition and cardiovascular disease in adult life. *Lancet, 341*, 938-941.

Bergman, A., Heindel, J. J., Kasten, T., Kidd, K. A., Jobling, S., Neira, M., Zoeller, R. T., Becher, G., Bjerregaard, P., Bornman, R., Brandt, I., Kortenkamp, A., Muir, D., Drisse, M. N., Ochieng, R., Skakkebaek, N. E., Byléhn, A. S., Iguchi, T., Toppari, J. and Woodruff, T. J. (2013). The impact of endocrine disruption: a consensus statement on the state of the science. *Environ Health Perspect, 121*, A104-106.

Bergman, Y. and Cedar, H. (2013). DNA methylation dynamics in health and disease. *Nat Struct Mol Biol, 20*, 274-281.

Beyersmann, D. and Hartwig, A. (2008). Carcinogenic metal compounds: recent insight into molecular and cellular mechanisms. *Arch Toxicol, 82*, 493-512.

Bolton, J. L., Smith, S. H., Huff, N. C., Gilmour, M. I., Foster, W. M., Auten, R. L. and Bilbo, S. D. (2012). Prenatal air pollution exposure induces neuroinflammation and predisposes offspring to weight gain in adulthood in a sex-specific manner. *FASEB J*, *26*, 4743-4754.

Byun, H. M., Benachour, N., Zalko, D., Frisardi, M. C., Colicino, E., Takser, L. and Baccarelli, A. A. (2015). Epigenetic effects of low perinatal doses of flame retardant BDE-47 on mitochondrial and nuclear genes in rat offspring. *Toxicology*, *328*, 152-159.

Castillo, P., Ibáñez, F., Guajardo, A., Llanos, M. N. and Ronco, A. M. (2012). Impact of cadmium exposure during pregnancy on hepatic glucocorticoid receptor methylation and expression in rat fetus. *PLoS One*, *7*, e44139.

Charalambous, M., da Rocha, S. T. and Ferguson-Smith, A. C. (2007). Genomic imprinting, growth control and the allocation of nutritional resources: consequences for postnatal life. *Curr Opin Endocrinol Diabetes Obes*, *14*, 3-12.

Chen, R. Z., Pettersson, U., Beard, C., Jackson-Grusby, L. and Jaenisch, R. (1998). DNA hypomethylation leads to elevated mutation rates. *Nature*, *395*, 89-93.

Crüts, B., van Etten, L., Törnqvist, H., Blomberg, A., Sandström, T., Mills, N. L. and Borm, P. J. (2008). Exposure to diesel exhaust induces changes in EEG in human volunteers. *Part Fibre Toxicol*, *5*, 4.

Cuozzo, C., Porcellini, A., Angrisano, T., Morano, A., Lee, B., Di Pardo, A., Messina, S., Iuliano, R., Fusco, A., Santillo, M. R., Muller, M. T., Chiariotti, L., Gottesman, M. E. and Avvedimento, E. V. (2007). DNA damage, homology-directed repair, and DNA methylation. *PLoS Genet*, *3*, e110.

Deaton, A. M. and Bird, A. (2011). CpG islands and the regulation of transcription. *Genes Dev*, *25*, 1010-1022.

Desaulniers, D., Xiao, G. H., Lian, H., Feng, Y. L., Zhu, J., Nakai, J. and Bowers, W. J. (2009). Effects of mixtures of polychlorinated biphenyls, methylmercury, and organochlorine pesticides on hepatic DNA methylation in prepubertal female Sprague-Dawley rats. *Int J Toxicol*, *28*, 294-307.

Donaldson, K., Tran, L., Jimenez, L. A., Duffin, R., Newby, D. E., Mills, N., MacNee, W. and Stone, V. (2005). Combustion-derived nanoparticles: a review of their toxicology following inhalation exposure. *Part Fibre Toxicol*, *2*, 10.

Dosunmu, R., Alashwal, H. and Zawia, N. H. (2012). Genome-wide expression and methylation profiling in the aged rodent brain due to early-life Pb exposure and its relevance to aging. *Mech Ageing Dev*, *133*, 435-443.

Dunn, G. A. and Bale, T. L. (2009). Maternal high-fat diet promotes body length increases and insulin insensitivity in second-generation mice. *Endocrinology, 150*, 4999-5009.

Foster, W. G. and Agzarian, J. (2008). Toward less confusing terminology in endocrine disruptor research. *J Toxicol Environ Health B Crit Rev, 11*, 152-161.

Fowler, P. A., Bellingham, M., Sinclair, K. D., Evans, N. P., Pocar, P., Fischer, B., Schaedlich, K., Schmidt, J. S., Amezaga, M. R., Bhattacharya, S., Rhind, S. M. and O'Shaughnessy, P. J. (2012). Impact of endocrine-disrupting compounds (EDCs) on female reproductive health. *Mol Cell Endocrinol, 355*, 231-239.

Gangisetty, O., Bekdash, R., Maglakelidze, G. and Sarkar, D. K. (2014). Fetal alcohol exposure alters proopiomelanocortin gene expression and hypothalamic-pituitary-adrenal axis function via increasing MeCP2 expression in the hypothalamus. *PLoS One, 9*, e113228.

Gilbert-Diamond, D., Cottingham, K. L., Gruber, J. F., Punshon, T., Sayarath, V., Gandolfi, A. J., Baker, E. R., Jackson, B. P., Folt, C. L. and Karagas, M. R. (2011). Rice consumption contributes to arsenic exposure in US women. *Proc Natl Acad Sci U S A, 108*, 20656-20660.

Godfrey, K. M., Sheppard, A., Gluckman, P. D., Lillycrop, K. A., Burdge, G. C., McLean, C., Rodford, J., Slater-Jefferies, J. L., Garratt, E., Crozier, S. R., Emerald, B. S., Gale, C. R., Inskip, H. M., Cooper, C. and Hanson, M. A. (2011). Epigenetic gene promoter methylation at birth is associated with child's later adiposity. *Diabetes, 60*, 1528-1534.

Green, B. B. and Marsit, C. J. (2015). Select Prenatal Environmental Exposures and Subsequent Alterations of Gene Specific and Repetitive Element DNA Methylation in Fetal Tissues. *Curr Environ Health Rep, 2*, 126-136.

Guo, J. U., Su, Y., Zhong, C., Ming, G. L. and Song, H. (2011). Hydroxylation of 5-methylcytosine by TET1 promotes active DNA demethylation in the adult brain. *Cell, 145*, 423-434.

Haddad, R., Kasneci, A., Mepham, K., Sebag, I. A. and Chalifour, L. E. (2013). Gestational exposure to diethylstilbestrol alters cardiac structure/function, protein expression and DNA methylation in adult male mice progeny. *Toxicol Appl Pharmacol, 266*, 27-37.

Haggarty, P., Hoad, G., Campbell, D. M., Horgan, G. W., Piyathilake, C. and McNeill, G. (2013). Folate in pregnancy and imprinted gene and repeat element methylation in the offspring. *Am J Clin Nutr, 97*, 94-99.

Hollingsworth, J. W., Maruoka, S., Boon, K., Garantziotis, S., Li, Z., Tomfohr, J., Bailey, N., Potts, E. N., Whitehead, G., Brass, D. M. and Schwartz, D. A. (2008). In utero supplementation with methyl donors enhances allergic airway disease in mice. *J Clin Invest*, *118*, 3462-3469.

Hoyme, H. E., May, P. A., Kalberg, W. O., Kodituwakku, P., Gossage, J. P., Trujillo, P. M., Buckley, D. G., Miller, J. H., Aragon, A. S., Khaole, N., Viljoen, D. L., Jones, K. L. and Robinson, L. K. (2005). A practical clinical approach to diagnosis of fetal alcohol spectrum disorders: clarification of the 1996 institute of medicine criteria. *Pediatrics*, *115*, 39-47.

Hoyo, C., Daltveit, A. K., Iversen, E., Benjamin-Neelon, S. E., Fuemmeler, B., Schildkraut, J., Murtha, A. P., Overcash, F., Vidal, A. C., Wang, F., Huang, Z., Kurtzberg, J., Seewaldt, V., Forman, M., Jirtle, R. L. and Murphy, S. K. (2014). Erythrocyte folate concentrations, CpG methylation at genomically imprinted domains, and birth weight in a multiethnic newborn cohort. *Epigenetics*, *9*, 1120-1130.

Huang, X. J., Shen, M., Wang, L., Yu, F., Wu, W. and Liu, H. L. (2015). Effects of tributyltin chloride on developing mouse oocytes and preimplantation embryos. *Microsc Microanal*, *21*, 358-367.

Ishii, D., Matsuzawa, D., Matsuda, S., Tomizawa, H., Sutoh, C. and Shimizu, E. (2014). Methyl donor-deficient diet during development can affect fear and anxiety in adulthood in C57BL/6J mice. *PLoS One*, *9*, e105750.

Järup, L. and Åkesson, A. (2009). Current status of cadmium as an environmental health problem. *Toxicol Appl Pharmacol*, *238*, 201-208.

Jacob, R. A., Gretz, D. M., Taylor, P. C., James, S. J., Pogribny, I. P., Miller, B. J., Henning, S. M. and Swendseid, M. E. (1998). Moderate folate depletion increases plasma homocysteine and decreases lymphocyte DNA methylation in postmenopausal women. *J Nutr*, *128*, 1204-1212.

Ji, Y. L., Wang, H., Liu, P., Zhao, X. F., Zhang, Y., Wang, Q., Zhang, H., Zhang, C., Duan, Z. H., Meng, C. and Xu, D. X. (2011). Effects of maternal cadmium exposure during late pregnant period on testicular steroidogenesis in male offspring. *Toxicol Lett*, *205*, 69-78.

Jiang, R., Jones, M. J., Sava, F., Kobor, M. S. and Carlsten, C. (2014). Short-term diesel exhaust inhalation in a controlled human crossover study is associated with changes in DNA methylation of circulating mononuclear cells in asthmatics. *Part Fibre Toxicol*, *11*, 71.

Johnston, H., Pojana, G., Zuin, S., Jacobsen, N. R., Møller, P., Loft, S., Semmler-Behnke, M., McGuiness, C., Balharry, D., Marcomini, A., Wallin, H., Kreyling, W., Donaldson, K., Tran, L. and Stone, V. (2013). Engineered nanomaterial risk. Lessons learnt from completed nanotoxicology studies: potential solutions to current and future challenges. *Crit Rev Toxicol*, *43*, 1-20.

Kafri, T., Ariel, M., Brandeis, M., Shemer, R., Urven, L., McCarrey, J., Cedar, H. and Razin, A. (1992). Developmental pattern of gene-specific DNA methylation in the mouse embryo and germ line. *Genes Dev*, *6*, 705-714.

Kašuba, V., Rozgaj, R., Milić, M., Želježić, D., Kopjar, N., Pizent, A. and Kljaković-Gašpić, Z. (2010). Evaluation of lead exposure in battery-manufacturing workers with focus on different biomarkers. *J Appl Toxicol*, *30*, 321-328.

Kelce, W. R., Monosson, E., Gamcsik, M. P., Laws, S. C. and Gray, L. E., Jr. (1994). Environmental hormone disruptors: evidence that vinclozolin developmental toxicity is mediated by antiandrogenic metabolites. *Toxicol Appl Pharmacol*, *126*, 276-285.

Kelsen, S. G., Duan, X., Ji, R., Perez, O., Liu, C. and Merali, S. (2008). Cigarette smoke induces an unfolded protein response in the human lung: a proteomic approach. *Am J Respir Cell Mol Biol*, *38*, 541-550.

Kilburn, K. H. (2000). Effects of diesel exhaust on neurobehavioral and pulmonary functions. *Arch Environ Health*, *55*, 11-17.

Kile, M. L., Houseman, E. A., Baccarelli, A. A., Quamruzzaman, Q., Rahman, M., Mostofa, G., Cardenas, A., Wright, R. O. and Christiani, D. C. (2014). Effect of prenatal arsenic exposure on DNA methylation and leukocyte subpopulations in cord blood. *Epigenetics*, *9*, 774-782.

Kimura, H. and Shiota, K. (2003). Methyl-CpG-binding protein, MeCP2, is a target molecule for maintenance DNA methyltransferase, Dnmt1. *J Biol Chem*, *278*, 4806-4812.

Kippler, M., Engström, K., Mlakar, S. J., Bottai, M., Ahmed, S., Hossain, M. B., Raqib, R., Vahter, M. and Broberg, K. (2013). Sex-specific effects of early life cadmium exposure on DNA methylation and implications for birth weight. *Epigenetics*, *8*, 494-503.

Kippler, M., Tofail, F., Gardner, R., Rahman, A., Hamadani, J. D., Bottai, M. and Vahter, M. (2012). Maternal cadmium exposure during pregnancy and size at birth: a prospective cohort study. *Environ Health Perspect*, *120*, 284-289.

Kitraki, E., Nalvarte, I., Alavian-Ghavanini, A. and Rüegg, J. (2015). Developmental exposure to bisphenol A alters expression and DNA methylation of Fkbp5, an important regulator of the stress response. *Mol Cell Endocrinol*, *417*, 191-199.

Koestler, D. C., Avissar-Whiting, M., Houseman, E. A., Karagas, M. R. and Marsit, C. J. (2013). Differential DNA methylation in umbilical cord blood of infants exposed to low levels of arsenic in utero. *Environ Health Perspect*, *121*, 971-977.

Kriaucionis, S. and Heintz, N. (2009). The nuclear DNA base 5-hydroxymethylcytosine is present in Purkinje neurons and the brain. *Science*, *324*, 929-930.

Kundakovic, M., Gudsnuk, K., Franks, B., Madrid, J., Miller, R. L., Perera, F. P. and Champagne, F. A. (2013). Sex-specific epigenetic disruption and behavioral changes following low-dose in utero bisphenol A exposure. *Proc Natl Acad Sci U S A*, *110*, 9956-9961.

Lee, K. W., Richmond, R., Hu, P., French, L., Shin, J., Bourdon, C., Reischl, E., Waldenberger, M., Zeilinger, S., Gaunt, T., McArdle, W., Ring, S., Woodward, G., Bouchard, L., Gaudet, D., Smith, G. D., Relton, C., Paus, T. and Pausova, Z. (2015). Prenatal exposure to maternal cigarette smoking and DNA methylation: epigenome-wide association in a discovery sample of adolescents and replication in an independent cohort at birth through 17 years of age. *Environ Health Perspect*, *123*, 193-199.

Lesseur, C., Armstrong, D. A., Paquette, A. G., Koestler, D. C., Padbury, J. F. and Marsit, C. J. (2013). Tissue-specific Leptin promoter DNA methylation is associated with maternal and infant perinatal factors. *Mol Cell Endocrinol*, *381*, 160-167.

Li, C., Guo, S., Gao, J., Guo, Y., Du, E., Lv, Z. and Zhang, B. (2015). Maternal high-zinc diet attenuates intestinal inflammation by reducing DNA methylation and elevating H3K9 acetylation in the A20 promoter of offspring chicks. *J Nutr Biochem*, *26*, 173-183.

Li, E., Bestor, T. H. and Jaenisch, R. (1992). Targeted mutation of the DNA methyltransferase gene results in embryonic lethality. *Cell*, *69*, 915-926.

Li, L., Zhang, T., Qin, X. S., Ge, W., Ma, H. G., Sun, L. L., Hou, Z. M., Chen, H., Chen, P., Qin, G. Q., Shen, W. and Zhang, X. F. (2014). Exposure to diethylhexyl phthalate (DEHP) results in a heritable modification of imprint genes DNA methylation in mouse oocytes. *Mol Biol Rep*, *41*, 1227-1235.

Li, X., Ito, M., Zhou, F., Youngson, N., Zuo, X., Leder, P. and Ferguson-Smith, A. C. (2008). A maternal-zygotic effect gene, Zfp57, maintains both maternal and paternal imprints. *Dev Cell*, *15*, 547-557.

Li, Y. J., Takizawa, H. and Kawada, T. (2010). Role of oxidative stresses induced by diesel exhaust particles in airway inflammation, allergy and asthma: their potential as a target of chemoprevention. *Inflamm Allergy Drug Targets*, *9*, 300-305.

Liang, P., Song, F., Ghosh, S., Morien, E., Qin, M., Mahmood, S., Fujiwara, K., Igarashi, J., Nagase, H. and Held, W. A. (2011). Genome-wide survey reveals dynamic widespread tissue-specific changes in DNA methylation during development. *BMC Genomics*, *12*, 231.

Liu, J., Ballaney, M., Al-alem, U., Quan, C., Jin, X., Perera, F., Chen, L.C. and Miller, R. L. (2008). Combined inhaled diesel exhaust particles and allergen exposure alter methylation of T helper genes and IgE production in vivo. *Toxicol Sci*, *102*, 76-81.

Liu, Y., Murphy, S. K., Murtha, A. P., Fuemmeler, B. F., Schildkraut, J., Huang, Z., Overcash, F., Kurtzberg, J., Jirtle, R., Iversen, E. S., Forman, M. R. and Hoyo, C. (2012). Depression in pregnancy, infant birth weight and DNA methylation of imprint regulatory elements. *Epigenetics*, *7*, 735-746.

MacNee, W. (2005). Pathogenesis of chronic obstructive pulmonary disease. *Proc Am Thorac Soc*, *2*, 258-266; discussion 290-251.

Madrigano, J., Baccarelli, A., Mittleman, M. A., Wright, R. O., Sparrow, D., Vokonas, P. S., Tarantini, L. and Schwartz, J. (2011). Prolonged exposure to particulate pollution, genes associated with glutathione pathways, and DNA methylation in a cohort of older men. *Environ Health Perspect*, *119*, 977-982.

Marjonen, H., Sierra, A., Nyman, A., Rogojin, V., Gröhn, O., Linden, A. M., Hautaniemi, S. and Kaminen-Ahola, N. (2015). Early maternal alcohol consumption alters hippocampal DNA methylation, gene expression and volume in a mouse model. *PLoS One*, *10*, e0124931.

Markunas, C. A., Xu, Z., Harlid, S., Wade, P. A., Lie, R. T., Taylor, J. A. and Wilcox, A. J. (2014). Identification of DNA methylation changes in newborns related to maternal smoking during pregnancy. *Environ Health Perspect*, *122*, 1147-1153.

Martinez-Arguelles, D. B. and Papadopoulos, V. (2015). Identification of hot spots of DNA methylation in the adult male adrenal in response to in utero exposure to the ubiquitous endocrine disruptor plasticizer di-(2-ethylhexyl) phthalate. *Endocrinology*, *156*, 124-133.

McCarrey, J. R. (2012). The epigenome as a target for heritable environmental disruptions of cellular function. *Mol Cell Endocrinol*, *354*, 9-15.

McMillen, I. C. and Robinson, J. S. (2005). Developmental origins of the metabolic syndrome: prediction, plasticity, and programming. *Physiol Rev*, *85*, 571-633.

Mortusewicz, O., Schermelleh, L., Walter, J., Cardoso, M. C. and Leonhardt, H. (2005). Recruitment of DNA methyltransferase I to DNA repair sites. *Proc Natl Acad Sci U S A*, *102*, 8905-8909.

Myers, G. J. and Davidson, P. W. (1998). Prenatal methylmercury exposure and children: neurologic, developmental, and behavioral research. *Environ Health Perspect*, *106 Suppl 3*, 841-847.

Nakamura, T., Arai, Y., Umehara, H., Masuhara, M., Kimura, T., Taniguchi, H., Sekimoto, T., Ikawa, M., Yoneda, Y., Okabe, M., Tanaka, S., Shiota, K. and Nakano, T. (2007). PGC7/Stella protects against DNA demethylation in early embryogenesis. *Nat Cell Biol*, *9*, 64-71.

Nel, A., Xia, T., Mädler, L. and Li, N. (2006). Toxic potential of materials at the nanolevel. *Science*, *311*, 622-627.

Nishijo, M., Satarug, S., Honda, R., Tsuritani, I. and Aoshima, K. (2004). The gender differences in health effects of environmental cadmium exposure and potential mechanisms. *Mol Cell Biochem*, *255*, 87-92.

Okai, Y., Sato, E. F., Higashi-Okai, K. and Inoue, M. (2004). Enhancing effect of the endocrine disruptor para-nonylphenol on the generation of reactive oxygen species in human blood neutrophils. *Environ Health Perspect*, *112*, 553-556.

Okano, M., Bell, D. W., Haber, D. A. and Li, E. (1999). DNA methyltransferases Dnmt3a and Dnmt3b are essential for de novo methylation and mammalian development. *Cell*, *99*, 247-257.

Oken, E., Levitan, E. B. and Gillman, M. W. (2008). Maternal smoking during pregnancy and child overweight: systematic review and meta-analysis. *Int J Obes (Lond)*, *32*, 201-210.

Olovnikov, I., Aravin, A. A. and Fejes Toth, K. (2012). Small RNA in the nucleus: the RNA-chromatin ping-pong. *Curr Opin Genet Dev*, *22*, 164-171.

Olsson, I. M., Bensryd, I., Lundh, T., Ottosson, H., Skerfving, S. and Oskarsson, A. (2002). Cadmium in blood and urine--impact of sex, age, dietary intake, iron status, and former smoking--association of renal effects. *Environ Health Perspect*, *110*, 1185-1190.

Onishchenko, N., Karpova, N., Sabri, F., Castren, E. and Ceccatelli, S. (2008). Long-lasting depression-like behavior and epigenetic changes of BDNF gene expression induced by perinatal exposure to methylmercury. *J Neurochem*, *106*, 1378-1387.

Ostro, B., Broadwin, R., Green, S., Feng, W. Y. and Lipsett, M. (2006). Fine particulate air pollution and mortality in nine California counties: results from CALFINE. *Environ Health Perspect*, *114*, 29-33.

Peters, S., Glass, D. C., Reid, A., de Klerk, N., Armstrong, B. K., Kellie, S., Ashton, L. J., Milne, E. and Fritschi, L. (2013). Parental occupational exposure to engine exhausts and childhood brain tumors. *Int J Cancer*, *132*, 2975-2979.

Pope, C. A., 3rd, Burnett, R. T., Thurston, G. D., Thun, M. J., Calle, E. E., Krewski, D. and Godleski, J. J. (2004). Cardiovascular mortality and long-term exposure to particulate air pollution: epidemiological evidence of general pathophysiological pathways of disease. *Circulation*, *109*, 71-77.

Power, C., Atherton, K. and Thomas, C. (2010). Maternal smoking in pregnancy, adult adiposity and other risk factors for cardiovascular disease. *Atherosclerosis*, *211*, 643-648.

Quenneville, S., Verde, G., Corsinotti, A., Kapopoulou, A., Jakobsson, J., Offner, S., Baglivo, I., Pedone, P. V., Grimaldi, G., Riccio, A. and Trono, D. (2011). In embryonic stem cells, ZFP57/KAP1 recognize a methylated hexanucleotide to affect chromatin and DNA methylation of imprinting control regions. *Mol Cell*, *44*, 361-372.

Roth, T. L., Lubin, F. D., Sodhi, M. and Kleinman, J. E. (2009). Epigenetic mechanisms in schizophrenia. *Biochim Biophys Acta*, *1790*, 869-877.

Ruchat, S. M., Houde, A. A., Voisin, G., St-Pierre, J., Perron, P., Baillargeon, J. P., Gaudet, D., Hivert, M. F., Brisson, D. and Bouchard, L. (2013). Gestational diabetes mellitus epigenetically affects genes predominantly involved in metabolic diseases. *Epigenetics*, *8*, 935-943.

Ruzov, A., Tsenkina, Y., Serio, A., Dudnakova, T., Fletcher, J., Bai, Y., Chebotareva, T., Pells, S., Hannoun, Z., Sullivan, G., Chandran, S., Hay, D. C., Bradley, M., Wilmut, I. and De Sousa, P. (2011). Lineage-specific distribution of high levels of genomic 5-hydroxymethylcytosine in mammalian development. *Cell Res*, *21*, 1332-1342.

Sakamoto, M., Yasutake, A., Domingo, J. L., Chan, H. M., Kubota, M. and Murata, K. (2013). Relationships between trace element concentrations in chorionic tissue of placenta and umbilical cord tissue: potential use as indicators for prenatal exposure. *Environ Int*, *60*, 106-111.

Satta, R., Maloku, E., Zhubi, A., Pibiri, F., Hajos, M., Costa, E. and Guidotti, A. (2008). Nicotine decreases DNA methyltransferase 1 expression and glutamic acid decarboxylase 67 promoter methylation in GABAergic interneurons. *Proc Natl Acad Sci U S A*, *105*, 16356-16361.

Schneider, J. S., Kidd, S. K. and Anderson, D. W. (2013). Influence of developmental lead exposure on expression of DNA methyltransferases and methyl cytosine-binding proteins in hippocampus. *Toxicol Lett*, *217*, 75-81.

Sen, A., Heredia, N., Senut, M. C., Land, S., Hollocher, K., Lu, X., Dereski, M. O. and Ruden, D. M. (2015). Multigenerational epigenetic inheritance in humans: DNA methylation changes associated with maternal exposure to lead can be transmitted to the grandchildren. *Sci Rep*, *5*, 14466.

Shen, L., Wu, H., Diep, D., Yamaguchi, S., D'Alessio, A. C., Fung, H. L., Zhang, K. and Zhang, Y. (2013). Genome-wide analysis reveals TET- and TDG-dependent 5-methylcytosine oxidation dynamics. *Cell*, *153*, 692-706.

Shivapurkar, N. and Poirier, L. A. (1983). Tissue levels of S-adenosylmethionine and S-adenosylhomocysteine in rats fed methyl-deficient, amino acid-defined diets for one to five weeks. *Carcinogenesis*, *4*, 1051-1057.

Skinner, M. K. (2014). Endocrine disruptor induction of epigenetic transgenerational inheritance of disease. *Mol Cell Endocrinol*, *398*, 4-12.

Smithells, R. W., Sheppard, S. and Schorah, C. J. (1976). Vitamin deficiencies and neural tube defects. *Arch Dis Child*, *51*, 944-950.

Somm, E., Stouder, C. and Paoloni-Giacobino, A. (2013). Effect of developmental dioxin exposure on methylation and expression of specific imprinted genes in mice. *Reprod Toxicol*, *35*, 150-155.

Susiarjo, M., Sasson, I., Mesaros, C. and Bartolomei, M. S. (2013). Bisphenol a exposure disrupts genomic imprinting in the mouse. *PLoS Genet*, *9*, e1003401.

Sutcliffe, J. S., Nelson, D. L., Zhang, F., Pieretti, M., Caskey, C. T., Saxe, D. and Warren, S. T. (1992). DNA methylation represses FMR-1 transcription in fragile X syndrome. *Hum Mol Genet*, *1*, 397-400.

Suzuki, T., Oshio, S., Iwata, M., Saburi, H., Odagiri, T., Udagawa, T., Sugawara, I., Umezawa, M. and Takeda, K. (2010). In utero exposure to a low concentration of diesel exhaust affects spontaneous locomotor activity and monoaminergic system in male mice. *Part Fibre Toxicol*, *7*, 7.

Syme, C., Abrahamowicz, M., Mahboubi, A., Leonard, G. T., Perron, M., Richer, L., Veillette, S., Gaudet, D., Paus, T. and Pausova, Z. (2010). Prenatal exposure to maternal cigarette smoking and accumulation of intra-abdominal fat during adolescence. *Obesity (Silver Spring)*, *18*, 1021-1025.

Tachibana, K., Takayanagi, K., Akimoto, A., Ueda, K., Shinkai, Y., Umezawa, M. and Takeda, K. (2015). Prenatal diesel exhaust exposure disrupts the DNA methylation profile in the brain of mouse offspring. *J Toxicol Sci*, *40*, 1-11.

Tahiliani, M., Koh, K. P., Shen, Y., Pastor, W. A., Bandukwala, H., Brudno, Y., Agarwal, S., Iyer, L. M., Liu, D. R., Aravind, L. and Rao, A. (2009). Conversion of 5-methylcytosine to 5-hydroxymethylcytosine in mammalian DNA by MLL partner TET1. *Science*, *324*, 930-935.

Tawa, R., Ono, T., Kurishita, A., Okada, S. and Hirose, S. (1990). Changes of DNA methylation level during pre- and postnatal periods in mice. *Differentiation*, *45*, 44-48.

Thirtamara Rajamani, K., Doherty-Lyons, S., Bolden, C., Willis, D., Hoffman, C., Zelikoff, J., Chen, L. C. and Gu, H. (2013). Prenatal and early-life exposure to high-level diesel exhaust particles leads to increased locomotor activity and repetitive behaviors in mice. *Autism Res*, *6*, 248-257.

Tsujiuchi, T., Tsutsumi, M., Sasaki, Y., Takahama, M. and Konishi, Y. (1999). Hypomethylation of CpG sites and c-myc gene overexpression in hepatocellular carcinomas, but not hyperplastic nodules, induced by a choline-deficient L-amino acid-defined diet in rats. *Jpn J Cancer Res*, *90*, 909-913.

Tucker, K. L. (2001). Methylated cytosine and the brain: a new base for neuroscience. *Neuron*, *30*, 649-652.

Valinluck, V., Tsai, H. H., Rogstad, D. K., Burdzy, A., Bird, A. and Sowers, L. C. (2004). Oxidative damage to methyl-CpG sequences inhibits the binding of the methyl-CpG binding domain (MBD) of methyl-CpG binding protein 2 (MeCP2). *Nucleic Acids Res*, *32*, 4100-4108.

Varrault, A., Gueydan, C., Delalbre, A., Bellmann, A., Houssami, S., Aknin, C., Severac, D., Chotard, L., Kahli, M., Le Digarcher, A., Pavlidis, P. and Journot, L. (2006). Zac1 regulates an imprinted gene network critically involved in the control of embryonic growth. *Dev Cell*, *11*, 711-722.

Waterland, R. A. and Michels, K. B. (2007). Epigenetic epidemiology of the developmental origins hypothesis. *Annu Rev Nutr*, *27*, 363-388.

Weaver, I. C., Champagne, F. A., Brown, S. E., Dymov, S., Sharma, S., Meaney, M. J. and Szyf, M. (2005). Reversal of maternal programming of stress responses in adult offspring through methyl supplementation: altering epigenetic marking later in life. *J Neurosci*, *25*, 11045-11054.

Wu, S., Zhu, J., Li, Y., Lin, T., Gan, L., Yuan, X., Xiong, J., Liu, X., Xu, M., Zhao, D., Ma, C., Li, X. and Wei, G. (2010). Dynamic epigenetic changes involved in testicular toxicity induced by di-2-(ethylhexyl) phthalate in mice. *Basic Clin Pharmacol Toxicol*, *106*, 118-123.

Xin, F., Jiang, L., Liu, X., Geng, C., Wang, W., Zhong, L., Yang, G. and Chen, M. (2014). Bisphenol A induces oxidative stress-associated DNA damage in INS-1 cells. *Mutat Res Genet Toxicol Environ Mutagen*, *769*, 29-33.

Yi, P., Melnyk, S., Pogribna, M., Pogribny, I. P., Hine, R. J. and James, S. J. (2000). Increase in plasma homocysteine associated with parallel increases in plasma S-adenosylhomocysteine and lymphocyte DNA hypomethylation. *J Biol Chem*, *275*, 29318-29323.

Zeisel, S. H. (2009). Importance of methyl donors during reproduction. *Am J Clin Nutr*, *89*, 673S-677S.

Zhang, L., Ding, S., Qiao, P., Dong, L., Yu, M., Wang, C., Zhang, M., Li, Y., Tang, N. and Chang, B. (2016). n-butylparaben induces male reproductive disorders via regulation of estradiol and estrogen receptors. *J Appl Toxicol*, (in press).

Zhang, X. F., Park, J. H., Choi, Y. J., Kang, M. H., Gurunathan, S. and Kim, J. H. (2015). Silver nanoparticles cause complications in pregnant mice. *Int J Nanomedicine*, *10*, 7057-7071.

Zhong, C. X. and Mass, M. J. (2001). Both hypomethylation and hypermethylation of DNA associated with arsenite exposure in cultures of human cells identified by methylation-sensitive arbitrarily-primed PCR. *Toxicol Lett*, *122*, 223-234.

Zoeller, R. T., Brown, T. R., Doan, L. L., Gore, A. C., Skakkebaek, N. E., Soto, A. M., Woodruff, T. J. and Vom Saal, F. S. (2012). Endocrine-disrupting chemicals and public health protection: a statement of principles from The Endocrine Society. *Endocrinology*, *153*, 4097-4110.

Zuo, X., Sheng, J., Lau, H. T., McDonald, C. M., Andrade, M., Cullen, D. E., Bell, F. T., Iacovino, M., Kyba, M., Xu, G. and Li, X. (2012). Zinc finger protein ZFP57 requires its co-factor to recruit DNA methyltransferases and maintains DNA methylation imprint in embryonic stem cells via its transcriptional repression domain. *J Biol Chem*, *287*, 2107-2118.

In: DNA Methylation
Editor: Kathleen Holland

ISBN: 978-1-53610-244-4
© 2016 Nova Science Publishers, Inc.

Chapter 2

CYTOSINE MODIFICATIONS AFFECT BIOLOGICAL SIGNIFICANCE BY A MULTIFACTORIAL NETWORK

Selcen Çelik-Uzuner[*], PhD
Karadeniz Technical University, Faculty of Science,
Department of Molecular Biology and Genetics, Trabzon, Turkey

ABSTRACT

Modifications at cytosine base are of great interests since they have been shown to alter during lifespan of many organisms and to involve in transcriptional regulation. The first discovered one, 5'-methylcytosine (5meC), can be converted to other modifications, 5'-hydroxymethylcytosine (5hmC), 5'-formylcytosine (5fC) and 5'-carboxycytosine (5caC) by enzymatic reactions, and the reversible conversions are also possible. This represents a model for epigenetic dynamism which is such represented by the hypothesis for active demethylation mediated by DNA repair. Some abnormalities associated with the changes in cytosine modifications suggest their importance for biological processes. Although there is a limited understanding of the biological importance of 5fC and 5caC, 5meC has been shown to associate with gene silencing, but 5hmC is commonly thought to involve in gene activation. Though this is not that simple since the effects of cytosine modifications can depend on such their genomic location, being within

---

[*] email: selcen.celik@ktu.edu.tr; phone number:+90 462 377 42 72.

cytosine-guanine repeats or not, the sequence of DNA part they are included, in association with three dimensional features of DNA structure indicating how cytosine modifications affect transcriptional outcome is hard to understand. This chapter presents transcriptional regulation by cytosine modifications through a broad concept of molecular sciences in physics, chemistry and biology, and suggests a complex network for cytosine modifications-mediated transcription depending on multiple factors.

Keywords: DNA methylation, modified DNA bases, methylome, gene regulation, transcriptional regulation

ABBREVIATIONS

3D	three dimensional
4meC	N4-methylcytosine
5caC	5'-carboxycytosine
5fC	5'-formylcytosine
5hmC	5'-hydroxymethylcytosine
5meC	5'-methylcytosine
6mA	N6-methyladenine
AID	activation induced deaminase
Alu	comes from the action of *Arthrobacter luteus* restriction endonuclease
APOBEC	apolipoprotein B mRNA editing enzyme, catalytic polypeptide-like
Arg	arginine
ATF4	activating transcription factor 4
BTB domain	broad-complex, tramtrack and bric a brac
C/EBPs	CCAAT-enhancer-binding proteins
CDKN	cyclin-dependent kinase inhibitor
CpG	cytosine-phosphate-guanine nucleotide
CTCF	CCCTC-binding factor
DAPK1	death associated protein kinase 1
DMRs	differentially methylated regions
DNMTs	DNA methyltransferase enzymes
ETS	E26 transformation-specific

H19	imprinted maternally expressed noncoding transcript
H3K14	histone H3 lysine 14
H3K14ac	histone H3 acetyl K14
H3K36me3	histone H3 tri methyl K36
H4K20me1	histone H4 mono methyl K20
HLA-G	human leukocyte antigen G
IGF-2	insulin-like growth factor 2
JUND	transcription factor JunD (a member of Jun protein family)
kj	kilojoule
KLF4	krüppel-like factor 4
KRAS	kirsten rat sarcoma viral oncogene homolog
LINE-1	long interspersed nuclear element 1
Lys	lysine
MBD	methyl binding domain
MBPs	methyl binding proteins
MeCP2	methyl CpG binding protein 2
mtDNA	mitochondrial DNA
Myc	myelocytomatosis oncogene
Nr3c1	nuclear receptor subfamily 3 group C member 1
NRF1	nuclear-respiratory factor 1
p53	tumour protein 53
PCNA	proliferating cell nuclear antigen
PDX1	pancreatic-duodenal homeobox factor 1
PPAR	peroxisome proliferator-activated receptors
Ras	rat sarcoma virus
SAM (AdoMet)	s-adenosylmethionine
Ser	serine
SMUG1	single-strand selective monofunctional uracil DNA glycosylase 1
Snrpn	small nucleoriboprotein n
SP1	specificity protein 1
SRA	set and ring finger-associated domain
Tat	tyrosine aminotransferase gene
Tcf3|Ascl1	transcription factor 3 | achaete-scute homolog 1
TDG	thymine DNA glycosylase

TES	transcription end site
TETs	ten-eleven translocation enzymes
TFs	transcription factors
TSS	transcription start site
Tyr	tyrosine
UHRF1	ubiquitin-like, containing PHD and RING finger domains, 1
UTR	untranslated region
Val	valine
Xist	X inactive specific transcript
ZBTB33	zinc finger and BTB domain containing 33
ZF2	zinc finger protein 2
ZF3	zinc finger protein 3
Zfp57	zinc finger protein 57 homolog

1. Introduction

The modifications of cytosine base within the genome of a range of organisms from bacteria to mammals have been a great interest of epigenetics in the field of molecular biology since these have been shown not to be only modifications but also involve in regulation of gene expression. The first discovered and the best known modification of cytosine is 5'-methylcytosine (5meC) which a methyl group is bound to the 5' carbon of cytosine predominantly within the CpG (Cytosine-phosphate-Guanine) dinucleotides (CpGs). Further modifications of 5meC catalysed by such TET (Ten-eleven translocation) enzymes include 5hmC (5'-hydroxymethylcytosine), 5fC (5'-formylcytosine) and 5caC (5'-carboxycytosine) [1-8] so that the term of "DNA methylation" currently represents all of these identified modifications of 5meC.

DNA methylation is involved in transcriptional regulation of genes during a range of cellular processes such as development [9, 10], X chromosome inactivation [11, 12], allele-specific expression (i.e., genomic imprinting) [13, 14] and tissue specific expression [15, 16]. 5meC pattern of genes or of whole genome can vary between healthy people, interestingly even if they are identical twins [17-22]. Methylation levels can also vary between different cell types from sample [23, 24]. Therefore, DNA methylation is a significant pattern specifying both intra- and inter-individual differences, and is very dynamic mechanism subject to change during lifetime.

The changes in 5meC have been also shown in a wide range of pathogeneses including cancer [25-28] and some neurological deficiencies [29-31]. These emerge two types of alterations occurred at whole genome and/or specific genes [32-34]. Increased methylation in promoters' of tumour suppressor genes, such of *p53* [35, 36] and *DAPK1* [37, 38], but decreased methylation in promoters' of proto-oncogenes, such *MYC* [39-41] and *RAS* [39, 42], is associated with cancers. The effect of methylation appear very important when the changes in DNA methylation occur at genes playing roles in critical processes that balance cell death or cell proliferation such tumour suppressor genes and proto-oncogenes.

A range of evidence supports that 5meC is associated with transcriptional repression [28, 34, 43, 44], but 5hmC has been shown to be involved in significantly increased gene expression [45-47] especially in nervous system [45, 46] consistent with its decrease found in brain with Alzheimer disease [48]. The decreased levels of 5fC and 5caC in patients with Alzheimer [29], similar to 5hmC, may indicate the role of these modifications specifically in nervous system. However, 5hmC has been shown to represses gene expression if it exists in gene promoters [49-51] but not in the gene bodies [49], and the effect of 5hmC at promoter regions differs between genes [52].

These show that cytosine modifications occurring as complex and dynamic mechanisms can regulate biological activities. Current understanding is based on biological significance of the changes in DNA methylation and the regulation of DNA methylation in different organisms [18, 22, 25-28, 53-55]. The well-known fact is that DNA methylation is highly associated with controlling transcription, in particular gene silencing. But, the further cytosine modifications are thought to be involved in transcriptional activation. At this point, the questions starting with "How" seem important. How do these chemical modifications affect genes? How do DNA modifications change DNA for the regulation of transcription? Regarding these kind of questions, this chapter focuses on the way(s) of affecting gene expression by the properties of cytosine modifications with the current knowledge. The story sounds a relevant combination of molecular sciences; physics, chemistry and biology.

2. STRUCTURES OF CYTOSINE MODIFICATIONS AND ACTIVE DEMETHYLATION PROCESS

The methyl group ($-CH_3$) is bound to carbon at 5' position of pyrimidine ring of cytosine (5'-methylcytosine) (Table 1A) in particular within the CpG dinucleotides, and was discovered a century ago [56] which has been then accepted as the fifth base of DNA molecule. DNA methyltransferases (DNMTs) are the responsible enzymes catalysing methylation of cytosines using the methyl donor, *S*-adenosylmethionine (SAM, AdoMet) [57] which is associated with the folic acid and vitamin B12 metabolisms [58]. Mammals have three main DNMTs; DNMT1, DNMT 2 and DNMT3A/3B. DNMT3A and DNMT3B are known to be responsible for *de novo* methylation of DNA which means the first establishment of methylation pattern of a cell during development such shortly after fertilisation [59, 60]. DNMT3L, DNMT3-like enzyme, differentially does not have a methyltransferase activity by itself but it regulates the activities of DNMT3A and DNMT3B [61]. The interaction of DNMT3L with DNMT3B is required for normal development of mouse [62]. However, DNMT1 plays a role for maintaining the pattern of DNA methylation established by DNMT3s during replication so that daughter cells have the same methylation pattern with mother cells at the end [63-65] (Figure 1A). This defines epigenetic inheritance. Recently, DNMT3A has been also shown to involve in DNA methylation maintenance with DNMT1, in particular neuron cells [66]. Differentially, DNMT2 enzyme is involved in the methylation of transfer RNA [67, 68] and further protein translation [69]. In zebrafish, there are more DNMTs described: Dnmt4, 5, 6, 7 and 8 which they are specifically involved in eye development [70].

More recently discovered metabolites of cytosine, 5hmC [3, 71], 5fC [8] and 5caC [5], are the current interests of epigenetics which have extended the context of "methylome." The functional groups of 5hmC, 5fC and 5caC bases are a hydroxyl (– OH), an aldehyde (– CHO) and a carboxyl (– COOH) group, respectively (Table 1A-B). These further modifications of 5meC catalysed by TET1,2,3 enzymes have been proposed to be produced in particular during the DNA repair pathway mediated by the loss of methylation (known as 'active demethylation') [72-76] (Figure 2). In this mechanism, DNA repair enzymes are also involved in the conversions of cytosine modifications, such TDG (thymine DNA glycosylase) is recruited in conversion of 5fC to C, of 5caC to C, and of 5hmC to C, AID/APOBEC (cytidine deaminases), MBD4 and SMUG1 enzymes can also involve in these reactions [4, 7, 73, 77-81]. However, it was also shown that AID/APOBEC did not have a high deamination activity on 5meC and also had no activity on 5hmC [82]. Besides, DNMT enzymes are able to catalyse the conversion of 5caC to unmodified cytosines directly [57]. The other mechanism for loss of DNA methylation is defined as "passive demethylation" which occurs during cell cycle. DNMT1 is responsible for the maintenance of DNA methylation so that methylates the complementary strand of hemi-methylated strand of DNA during replication. If the activity of DNMT1 is blocked by nucleoprotein complexes (Figure 1B) or inhibited by histone acetylation (Figure 1C) so that it fails to methylate the unmethylated strand, consequently DNA methylation in daughter cells cannot be maintained and methylation pattern is lost [83].

The possible conversion of cytosine modifications to each other that represents a reversible and therefore dynamic mechanism regulated by the enzymatic activities makes scientists attracted to reveal the effects of these on biological models and to improve therapeutic approaches for epigenetic-related diseases. This chapter mainly focuses on the modifications occurring at the carbon in the fifth position of cytosine ring, but there is also another cytosine modification formed by the addition of methyl group to nitrogen atom bound to fourth carbon, called 4meC, in the bacterial genome [84] (also mentioned below).

Table 1. Chemical properties of cytosine modifications

Modification	Structure (A)	Functional group (B)	Chemical bonds (C)
5meC		Cyt - CH3 Hydrophobic	C - C (*) C - H
5hmC		Cyt - OH Hydrophilic	C - O (*) O - H
5fC		Cyt - CHO Hydrophilic	C - C (*) C = O C - H
5caC		Cyt - COOH Hydrophilic	C - C (*) C = O C - O O - H

Cyt: cytosine
(*) Bond between functional group and cytosine

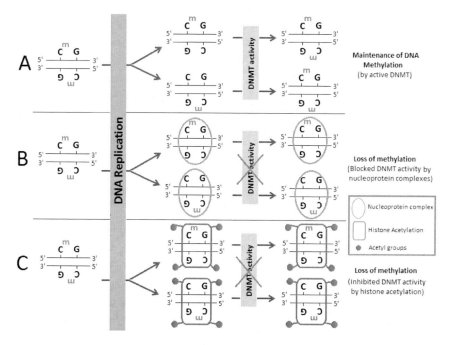

Figure 1. Passive Demethylation in DNA replication. In DNA replication, two methylated DNA strands of the genome are separated into daughter chromatids and form hemimethylated DNA strands, followed by either that (A) in normal circumstances, DNMT completes the symmetric methylation, and therefore maintains DNA methylation pattern of the original strand, or DNMT activity is blocked either by (B) regulatory nucleoprotein complexes or (C) histone acetylation, resulting in the loss of methylation. Adapted from Wolffe et al. (1999) [85].

The structures of cytosine modifications define DNA-protein interactions by inducing physical and chemical changes in DNA. The interaction between these changes provides a complicated mechanism for biological regulation by cytosine modifications. To focus on this mechanism, this chapter includes physical and chemical alterations associated with DNA methylation and their possible effects on transcriptional regulation.

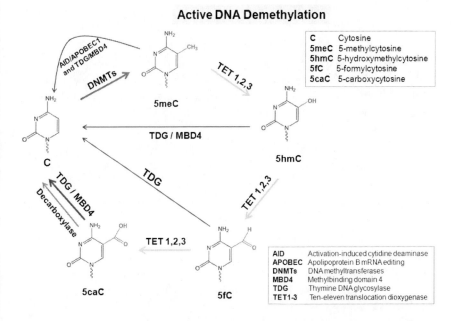

Figure 2. Current model for active DNA demethylation. This mechanism includes the removal of methylation from cytosine followed by hydroxylation (5hmC), formylation (5fC) and carboxylation of cytosine (5caC) (Those structures are given in Table 1). Involved enzymes are TETs, TDG, MBD4, decarboxylases and AID/APOBEC1, and base excision repair system also involved in demethylation process to repair DNA. DNMTs methylate cytosines after DNA repair's completed, and original methylation pattern is therefore maintained.

3. CLASSIFICATION OF DNA METHYLATION IN TERMS OF GENETIC MATERIALS AND GENOMIC REGIONS

3.1. Nucleic Acid Type

Regarding the existence of cytosine in each type of genetic material, it is expected to find cytosines methylated not only in nuclear DNA but also in mitochondrial DNA (mtDNA) [86, 87], plastid DNA [88] and RNA [89, 90]. These are not such a classification of 5meC (and its derivatives) location only, but also expected to be important in biological manner. These methylations are detailed below.

3.1.1. Nuclear DNA Methylation

Cytosine methylation mainly occurs in the nuclear DNA. DNA within the nucleus presents the majority of the genetic information of the organism. In eukaryotes, nuclear DNA is packaged with chromatin proteins, such histones, and thus exists in the nucleus in a compact form. Nuclear DNA methylation is highly associated with chromatin structure and its reorganisation during cellular processes. The pattern of nuclear DNA methylation plays roles in (i) early development, (ii) X chromosome inactivation, (iii) genomic imprinting and (iv) tissue-specific gene expression (Figure 3). DNA methylation is involved in these processes by inducing gene inactivation resulted from inaccessible docking sites for transcription factors (Figure 4). The common abnormalities associated with the alterations of DNA methylation are cancers, neurodegenerative diseases and developmental abnormalities (Figure 3).

(i) In early development, shortly after fertilisation, paternal genome is rapidly demethylated prior to DNA replication in mice [91-93] and in rabbit [94]. However, the methylation mark of the maternal genome is maintained over several cell cycles [91-94]. In further progress, such in 2-cell embryos, the level of DNA methylation is similar to zygote [95].

(ii) X chromosome inactivation occurs during preimplantation development within the female genome in the cells of inner cell mass at embryonic day 3.5 to balance X-related genes, and which X chromosome is inactivated is randomly determined [96]. *Xist* gene is the only gene which is expressed in inactive X chromosome of the female genome. Although CpG loci near the 5' end of *Xist* gene promoter are not methylated in the murine sperm, mature murine oocytes are methylated. Maternal-based methylation at *Xist* gene is also sustained in the male embryo. It suggests that imprinting pattern of *Xist* is transmitted to next generations by maternal-derived methylation [97].

(iii) Genomic imprinting is described as the mechanism of mono-allelic silencing of gene inherited from either mother or father. This process is represented by such differentially methylated regions (DMRs) of the genome that have different methylation patterns between female and male alleles of the same cell. For instance, maternal allele of *H19* gene is activated, however paternal allele of *Igf-2* is repressed by DMRs [98]. *Snrpn* gene is methylated in oocytes only, and its methylation pattern did not change during somatic cell differentiation [99], and it is monoallelically methylated before gametogenesis in mice [100].

(iv) DNA methylation also functions in tissue-specific gene expression by three possible ways: (1) DNA methylation level can decrease (called demethylation or hypomethylation) or (2) DNA methylation level can increase (hypermethylation) or (3) DNA methylation can be targeted by some transcription activator proteins. (1) In the first case, for instance, DNA methylation is lost or at very low levels in some CpGs in testis, but found at remarkable levels in kidney and liver [101], and another example includes the demethylation of *INSULIN* and *PDX1* genes in pancreatic beta cells [102] therefore recruits increased expression of these genes working only in beta cells of pancreas. (2) In contrast, these beta cell-specific genes are methylated in fibroblasts which do not function in insulin metabolism [102]. (3) DNA methylation can be also targeted by some transcription activators which have a binding affinity to methylated DNA, such C/EBP proteins (details in Section 6.4). This can be observed in specific genes involved in keratinocyte differentiation [103].

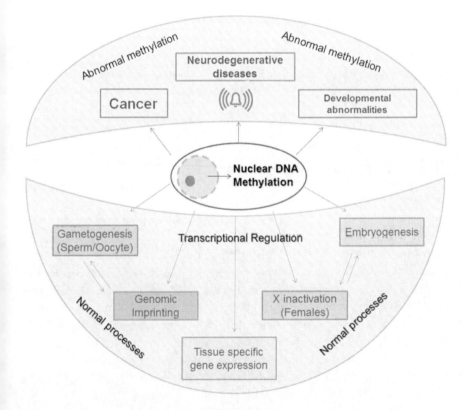

Figure 3. Functions of nuclear DNA methylation in normal processes of cells and abnormalities associated with DNA methylation alterations.

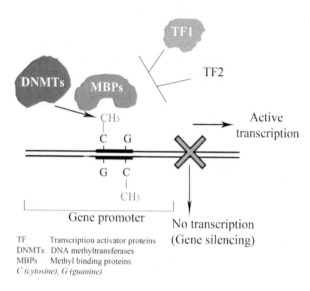

Figure 4. Transcriptional silencing by CpG methylation. The figure represents one possible way for transcriptional silencing by CpG methylation associated with methyl-binding proteins and DNA methyltransferase enzymes. This protein complex blocks the binding of transcription factors to the specific site of the genome, and results in gene inactivation. Simplified from Burgers et al. (2002) [104].

3.1.2. Organelle DNA Methylation

The organelles of eukaryotes that consist of their own DNA molecules are mitochondria and plastids. These are in circular formation differentially than linear nuclear DNA, and they are not packaged. But, cytosines in the organelle DNAs can be also methylated similar to the nuclear DNA. For instance, mitochondrial DNA (mtDNA) methylation has been shown in such human fibroblasts [86, 87], and has its own epigenetic mark differentially driven than nuclear DNA, but in association with nuclear DNA within the same cell [105, 106]. mtDNA methylation was also present with a characteristic level in germ cells but it was lost in early development [10]. Although mtDNA methylation was rarely detectable in human gastrointestinal cancer cells [107], it appears as an important epigenetic mark for biological regulations because of its alteration found in colon cancer [108], and a liver disease [109]. These suggest that mtDNA methylation pattern can vary depending on tissue type and developmental stage of the organism. mtDNA exists in 1000 copies in such a mouse fibroblast cell [87, 110], and each mitochondrion can have 2-10 copies of its own DNA [111, 112]. This can suggest that mtDNA methylation forms a significant level of cellular methylation. Cytosine methylation is not the only modification present in mtDNA but 5hmC has been also shown to exist [113]. The level of 5hmC in mtDNA diminished in frontal cortex but not the cerebellum during aging, whereas this was not the case for mitochondrial 5meC [114]. This can indicate a specific role of mitochondrial 5hmC as well as nuclear 5hmC in nervous system.

Similar to mtDNA, cytosine methylation exists in plastid genomes. But, interestingly it is not associated with gene expression as shown in chloroplast DNA of *Nicotiana tabacum* [88]. Chloroplast DNA has been used as an internal control for detection of genome-wide DNA methylation because it is usually unmethylated [115]. However, inactivation of some photosynthesis genes regulated by cytosine methylation was shown to mediate transformation of chloroplasts to chromoplasts in tomato [116].

3.1.3. RNA Methylation

Cytosines in the RNA molecule are methylated by DNMT2 enzyme and this methylation is involved in protein synthesis [117]. RNA itself can also mediate DNA methylation in plants and this mechanism functions such in protection against pathogens and in cellular response to stress [118]. 5hmC modification is also found in RNA and it is catalysed by TET enzymes as well as in DNA, and the existence of 5fC modification in RNA molecule has been recently revealed [119-121]. However, their roles in cellular progress are not clear yet.

3.2. Cytosine-Guanine (CpG) Repeats

3.2.1. CpG Content

Genes are classified according to their CpG content into three: CG-poor, CG-medium and CG-rich genes. CpG repeats tend to be methylated more around nucleosomes (with 0-100bp distance from nucleosomes) in both human and *Arabidopsis* [122]. CG content is important since many of genes have remarkable CG content at promoter regions [123] but with different scale such as housekeeping genes with a high content of CpG [124-126].

3.2.2. CpG Islands

CpG intensity along the genome is classified as (a) CpG islands, (b) shore regions (<2kb flanking CpG islands), (c) shelf regions (<2kb flanking outwards from a CpG shore) and (d) non-CpG regions. The methylation levels were detected high in CpG islands of shore and shelve regions [127]. Since each cytosine base potentially can be methylated so that this is not surprising to determine cytosines methylated within non-CpG regions of the genome as well [128, 129]. This is also the case for mtDNA as non-CpGs in the promoter of its D-loop region have been shown to be methylated in mammals [113].

3.3. Intergenic Locations

CpGs can be found in gene promoters [124, 126, 130], in gene bodies [127, 131, 132], in introns [133-136] and in exons [134, 136]. Exons with a high content of CG than flanking introns have a higher nucleosome settlement than the flanking introns consistent with the more methylation level found in exons than introns [134]. In addition to the regions within a gene (inter-genic regions), distribution of DNA methylation is highly wide along regions between genes (intra-genic regions) [24, 137]. DNA methylation occurring in intra-genic regions plays a role for transcriptional regulation of various genes [138].

4. 3D DNA STRUCTURE AND INTERACTION WITH METHYLATION

The physical characteristics, CG base pair content of genes/genomes and the location of CpGs, are related to morphological factors within physical characteristics of DNA. However, there are some other features more related to molecular structures in physical characteristics of DNA. These properties of DNA include (i) its double-strand helix structure, (ii) its sequence (original base pairs and/or mutated bases and/or modified bases), and (iii) its packaging with chromatin proteins, i.e., histones, to form nucleosomes or its releasing from bound chromatin proteins.

(i) The most common form of DNA double-helix, B-form, has 10.1 bases per turn and consists of two characteristic major and minor grooves, and has been defined as an average structure of DNA by Watson and Crick [139]. Both grooves have almost the same deep, but major groove is 2-fold wider than of minor groove within B-form [140]. The major groove provides docking sites for proteins to bind through their α-helices [141] while the narrower minor groove is generally recognised by β-sheet structures of TATA box-binding proteins [141, 142]. But, as an alternative mode, some proteins can bind to DNA using α-helical and β-sheet structures to bind minor and major grooves, respectively, [143, 144], and this is also possible that many of transcription factors can bind DNA via different contacts at the same time [145]. This complexity of DNA-protein interaction further includes some proteins that bind to DNA regardless the specific base pairs of DNA [146], but some of DNA binding proteins that can selectively recognise base pairs [141, 144, 147] (base-readout), or specific DNA sequences (shape-readout) [141, 145, 147, 148]. Shape-readout refers both local (e.g.., narrowing the minor groove) and global morphology (e.g., bending of DNA helix) [149]. Methyl group bond to cytosine lies on the major groove of DNA double-strand helix [150] due to methyl binding the 5' position of cytosine (Figure 5A). Therefore, it is not surprising that formylation of cytosines (5fC) can also occur in CpG repeats [151, 152] lying in the major groove [151] since 5meC and 5hmC are the precursors of 5fC. As a result of formylation of cytosines at CpG repeats, the major groove is tightened but the minor groove is enlarged [151]. The helix structure is subject to alter as 5fC induced CpG-containing oligomers to right-handed helix structure [152]. The high density of 5fC is correlated to the ellipticity of DNA as it prompts helical under-winding of DNA during supercoiling (called F-DNA) [151]. Conformational changes induced by DNA methylation highly depend on the specific DNA sequences. For instance, a different DNA structure has been shown to be induced by cytosine methylation *(underlined)* of d(GGCGCC)2 sequence, and described as E-DNA [150]. However, methylation *(underlined)* at the sequence d(CCAGGCCTGG) did not have an effect on DNA conformational changes [153]. These suggest that cytosine modifications can be a factor for the changes in 3D DNA helix conformation but depending on DNA sequence in some cases.

DNA-helix structure is defined by (ii) the specific base pairs/sequences which stimulate shaping the helix. For example, DNA helix bending of major groove preferentially happens at the GC pair especially in GGCC sequence [154, 155]. AT pairs or AAAA sequences narrow the minor groove, but GC pairs broaden it [149]. Major groove of B-DNA structure is narrowed by a roll bend at AT/GC junction but this bending is occasionally formed depending on the DNA-protein interactions [156]. These suggest that DNA sequence is not important for genetic code only, but also for spatial organisation of the genomic information. This dynamic shaping can be supposed to affect the formation of nucleosomes. Nucleosome components involve in packaging a 2-3 meters long human DNA molecule to fit into a nucleus. At this point, calling DNA as "nucleosomal DNA" seems more appropriate and inclusive. (iii) Nucleosomal DNA is involved in the mechanism of DNA methylation, and the relationship between DNA methylation and nucleosome structure is interactive. Cytosine methylation can affect the nucleosome arrangement [157, 158] or *vice versa* [122]. For instance, histone acetylation (at H3K9 and H3K14) is associated with the CpG content of the promoter regions of the genes, but H3K14 is correlated with inactive promoters relatively [159]. But histone methylation (at H3K9 and H3K36me3) is associated with cytosine methylation in gene bodies [132], and there is a positive correlation between H3K4 trimethylation and the CpG content [160].

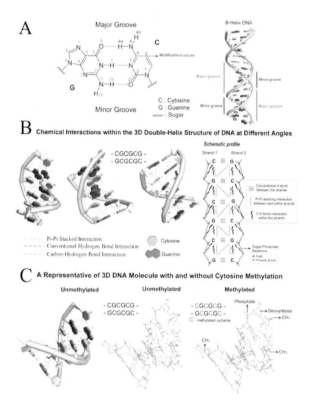

Figure 5. Overview of 3D DNA Molecule and CG Methylation. (A) Standard IUPAC numbering for cytosine-guanine base pair (left) and common B-DNA helix structure with minor and major grooves (right) are given. Cytosine modifications at 5' carbon lie on major groove of DNA. (B) Non-covalent interactions within the DNA molecule, pi-pi stacking, hydrogen bond interactions are shown at different points of view. The model DNA sequence shown is CGCGCG with 5' carbons indicated. A schematic profile of these interactions between and within the DNA strands is provided on sugar-phosphate backbone (right). (C) A representative methylated DNA model is given with 5' carbons of cytosines indicated. *(This figure was created using the software, Discovery Studio V4.1 by Biovia, CA, USA).*

Cytosine methylation can affect the mechanical stability and electromechanical properties of DNA molecule [161, 162]. 5hmC modification can reverse methylation-mediated DNA stability, but it is still able to stabilise DNA helix relatively lower than 5meC's [152], and was shown to increase stability of DNA in zipper geometry but less than shear one [163]. However, this is not the case for 5fC and 5caC. 5fC and 5caC showed a lower melting temperature (around 54°C) similar to unmodified cytosine than both 5meC and 5hmC that resulted in higher melting temperatures required (60°C and 58°C, respectively) [151]. Each cytosine (with any way of modification) always contributes to the double-strand structure by pairing with guanine of the complementary DNA strand unless a mutation exists. However, each modification can differentially affect the DNA helix stability which defined by hydrogen bonding between strands (between cytosine and guanine in this case) suggesting that cytosine modifications provide distinctive properties to DNA stability. Together, such these physical properties appear significant as these can induce electrostatic status [141, 164], DNA bending [165] and binding kinetics [166, 167]. The numerical changes at distances and angles between atoms can contribute on these so that both physical and geometrical or mathematical properties describe 3D architecture of genetic material, and cytosine modifications can interact with these properties.

5. CHEMICAL CHARACTERISTICS OF DNA AND ITS RELATION WITH METHYLATION

Beyond the alterations of physical/mathematical properties of 3D DNA molecule, chemical characteristics including (i) chemical bonds, (ii) functional groups, and (iii) properties of binding proteins can also determine and affect the nature of DNA structure. (i) For instance, methyl group itself has only C – H bond and is bound to cytosine via a C – C bond, but hydroxymethylation provides an O – H bond and connects to cytosine via the C – O bond. The number of chemical bond type increases with further conversions, as formylation adds C – H and C – O, and carboxylation adds C – O, C = O and O – H bonds (Table 1C). As well as methyl group, both formylation and carboxylation provide C – C bonds for binding to the cytosine's carbon at 5th position, but the hydroxyl group binds to carbon via C – O bond. C – C single bond is the weakest one of single covalent bonds [168, 169]. C – C bond energy (347kj/mole) is almost the same with C – O bond energy (350kj/mole) [169,

170] suggesting that each of these functional groups modifying cytosine has similar stability to undergo chemical reactions as the bond energies present the stability of molecules [170]. This can also support that the possible conversion of each modification to its downstream metabolite (*i.e.*, 5meC to 5hmC, 5hmC to 5fC, and 5fC to 5caC) and upstream conversion of each to unmodified cytosine are reasonable. The C – C bond cleavage rate of 5fC and 5caC is more than 5hmC [171]. This means that 5fC and 5caC can be removed faster than 5hmC to result in the unmodified cytosine. Methylated cytosines also increased the rate of *DNase* cleavage to some extent [172] suggesting that DNA methylation may affect DNA-protein interactions since DNA is protected from enzymatic cleavage in case of proteins bound to DNA. Much earlier, DNA cleavage by a restriction endonuclease was however shown to be reduced by site-specific DNA methylation in canonical DNA sequences (sequences accepted as the most commonly occur) [173]. SAM, the substrate for methylation catalysed by DNMTs, is also one of the cofactors for restriction enzymes during DNA cleavage [174, 175] which is importantly involved in regulation of 3D DNA structure and supercoiling DNA by such topoisomerases, hence in DNA-protein interactions. These imply a potential effect of methylation on phosphate-sugar backbone structure of DNA helix. These together can mean that chemical bonds provided by chemical groups allow DNA undergoing chemical reactions at different rates.

In 3D DNA structure, there are possible non-covalent interactions between and within the DNA strands suggesting the complex organisation of DNA itself. These include (a) the interaction occurring between two strands only (i.e., conventional hydrogen bond interaction between the complementary bases), (b) the interaction occurring within strands only (i.e., carbon-hydrogen bond interaction between and within the bases), and (c) the interaction occurring both between and within the strands (i.e., pi–pi (π–π) stacking interaction occurring between two complementary and cytosine, and the same DNA bases, such guanine, through their aromatic rings) (Figure 5B-C). DNA stair motif also includes the cation – π interaction which is another non-covalent interaction between aromatic rings and cations such Na$^+$ [176, 177]. Proteins can also use the cation – π interaction to complex with DNA. For instance, the catalytic domain of human TET2 enzyme binds to modified cytosines by the sandwich structure of its aromatic residue (phenylalanine, tyrosine or tryptophan) and its cation residue (arginine or histidine) [178]. Methylation provides a CH-π interaction, which is considered as a weak force arising among CH groups and π structures, between cytosine and flanking thymine [179]. CH – π interaction is a kind of hyperconjugation interaction which is known to affect e.g., stability and bond lengths [180], and these interactions between the aromatic rings and cations have also electrostatic properties so that these can be also included in the physical properties of DNA.

Beyond the complexity of DNA itself, **(ii)** addition/changing of chemical group to DNA is principally expected to contribute new characteristics to the molecule, therefore affects the relations of DNA within its bio-network such with proteins. One of these prominent characteristics is polarity. For instance, methyl group ($-CH_3$) is an aliphatic compound classified in alkyl groups and has hydrophobic (non-polar) characteristics (C – H bond generally considered as hydrophobic) whereas metabolites of methylated cytosine have functional groups which are hydrophilic (polar) (Table 1B). In regard to the polar functional groups, amides have the highest polarity followed by acids, alcohols (5hmC), amines, ketones, aldehydes (5fC), esters, ethers and alkanes, respectively. Consistently, 5meC is more hydrophobic than 5hmC [82]. Accordingly, **(iii)** each chemical group can define and affect the relationship between the DNA molecule and proteins. There are two major chemical mechanisms for DNA binding specificity of proteins. One is regulated by hydrophobic contacts of proteins with DNA bases on the major groove. The other and more common mechanism for recognition of DNA by proteins is composed of hydrogen bond (H-bond) formation in the major groove [149]. Hydrophobic interaction can involve in DNA packaging with proteins in particular the association of methylated DNA with chromatin proteins, *i.e.*, histones that contact to where each turn of DNA helix via its arginine side chains [164], therefore it is crucial for both structuring complex organisation of genetic material within the nucleus [157, 158] and the association of cytosine methylation with heterochromatin which is a tightly packaged inactive chromatin section [181, 182]. In contrast, 5hmC highly accumulated in euchromatic regions not in heterochromatin [183, 184] suggesting the role of 5hmC for active gene expression.

6. PROTEINS BINDING TO METHYLATED DNA

6.1. Methylated DNA-Specific Proteins

There are three major methyl-binding protein (MBPs) groups with high affinity for binding to methylated DNA, via a specific domain recognising methylated cytosines: (6.1.1) MBD proteins which contain a common methyl-binding domain (MBD), (6.1.2) Kaiso-like zinc finger proteins and (6.1.3) Set and Ring finger-associated (SRA) domain proteins [185-187] (Figure 6).

6.1.1. MBD Proteins

MBD proteins have a methyl-binding domain (MBD) and recognise and bind to methylated DNA by these domains. Identified MBD proteins are MBD1, MBD2, MeCP2, MBD3, MBD4, MBD5 and MBD6. MBD1 has the highest binding affinity (90-fold) to methylated DNA over unmodified cytosine. This is followed by MBD2 (35-fold), MBD4 (11-fold), and MeCP2 (8-fold), but interestingly MBD3 does not have a binding affinity to methylated DNA [188], but to hydroxymethylated DNA [189]. MBD5 and MBD6 were shown to not to bind to methylated DNA in a sequence-dependent context in human [190]. MBDs interact with methylated DNA through the major groove because methyl groups lies on the major groove.

A way of the contact between MBDs and DNA includes that MBD domain of MBPs can distinguish the methylated DNA by their arginine residues through the interaction between hydrogen bonding and cation – π interface because methylation of cytosines induced the hydrophobic surface connecting MBD and methylated DNA [187]. MBD proteins were shown to have Val20, Arg22, Tyr34, Arg44, Ser45 residues [187, 191]. Such serine (Ser) and valine (Val) can make hydrophobic contacts, but arginine (Arg) and tyrosine (Tyr) residues form both hydrophobic interaction with methylated cytosines and hydrophilic interaction with hydrogen bond between two DNA strands [191]. Mutations at Arg22 and Arg44 of MBD1 caused a decrease of the contact area of methylated DNA and therefore, unsurprisingly, reduce the methylated CG affinity [187]. The most conserved regions of MBD domain lie between Arg22 and Lys46 including loop L1, β3, β4 strands and loop L2 which are responsible for recognition of methylated DNA [187]. Arginine in MeCP2 and MBD2 proteins forms hydrogen bond and stacking with the methylated cytosines [192, 193]. Arg22 and Arg44 placing in major groove of DNA result in a formation of hydrogen bond with guanine and bending guanines to minor groove of DNA [187]. Arginine facilitates binding of MBD proteins to hydrophobic methylated DNA since arginine is able to decrease the hydrophobic interaction [194] (Figure 7A) because it is considered as the most hydrophilic amino acid with – 1.01 of hydrophobicity scale according to the algorithm discovered by Fauchere and Pliska [195]. Methyl group makes DNA hydrophobic resulting in the limitation of binding of some proteins to methylated DNA. Therefore, arginine is an ideal amino acid in such MBD proteins to be used for binding to DNA. It can be concluded that at the same time, arginine may also involve in maintenance the hydrogen bond between two DNA strands in the DNA site which MBDs bind.

The close relationship of MBD proteins with the existence of methylation is consistent with such that MBD1 involving in transcriptional repression regulated by DNA methylation [196] within the heterochromatin [197, 198]. However, MBD1 protein is also able to accumulate within the euchromatic regions of genome but still associated with gene inactivation by methylation [199]. This should be also noticed that MBD of MeCP2 is specific for binding to the methylated DNA but has not [200] or a low affinity to the hydroxymethylated DNA in the existence of salt [201] suggesting the importance of ions for DNA-protein interaction.

MBD1 protein can also have a CXXC domain which can facilitate to binding with the low levels of methylation [202] or unmethylated CpGs [203] supporting the findings showed that MBD1 protein acted independent from methylated cytosines to some extent [204, 205], and other members, MBD2 [203], MBD3 [206], MBD5 and MBD6 [190] are also able to bind unmethylated DNA. These can suggest the different role(s) of MBD proteins in biological activities rather than in DNA methylation machinery only.

6.1.2. Kaiso-Like Zinc Finger Protein Group

Kaiso can bind to methylated DNA without having a methyl binding domain, in contrast to MBD proteins [207], and Kaiso-like proteins also do not contain a methyl binding domain [185]. Kaiso group is defined by the use of its own helix-turn-helix domain to form hydrogen bonds with bases in the major groove, and it has an affinity for binding to the methylated DNA via its ZnF (Zinc finger) domain. The prominent one of these proteins is Kaiso (also known as ZBTB33), and similar to the MBD proteins, it contacts with the methylated DNA by its arginine [208] (Figure 7B). However, *in vivo* studies showed binding of Kaiso to unmethylated promoters of genes [209]. Kaiso may have a dual function both in transcriptional activation and repression. For instance, in cancer, Kaiso was found to bind to the *CDKN2A* gene (functioning in DNA damage response), in a methylation-dependent manner [210] suggesting its involvement in silencing tumour suppressor genes by methylation. However, it has a role in p53-controlled cell death through increasing the binding of p53 protein to *CDKN1A* gene's promoter [211]. Krüppel-like factor 4 (KLF4), encoded by a pluripotency gene (*KLF4*), also has a zinc-finger protein structure and can bind to the methylated DNA [208, 212-215]. KLF4 has been shown to function in the genomic stability [216], and have an anti-tumour activity [217, 218]. CTCF protein, another zing finger protein, has been recently shown to implicate in the conversion of 5meC to 5hmC via TET enzymes and to have a role for transcriptional activation [219]. Zfp57 protein makes a hydrophobic contact via its Arg178 [220] and it, as well as the examples above, suggests the definitive role of arginine for recognising methylated DNA.

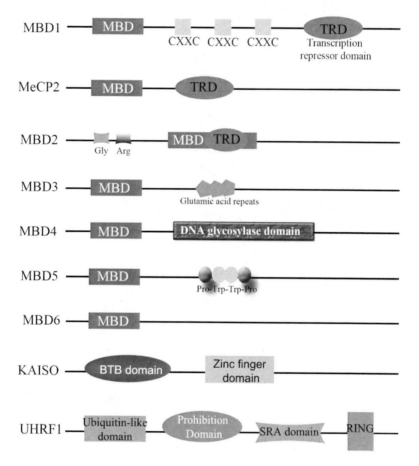

Figure 6. The schematised basic structures of methylated DNA-specific proteins. This figure shows the basic structures of MBD proteins (MBD1-6 and MeCP2), Kaiso and UHRF1 (a SRA protein) with specific regions.

6.1.3. Set and Ring Finger-Associated (SRA) Domain Proteins

Similar to Kaiso proteins, SRA domain proteins do not contain a MBD [185]. Some members of these proteins are UHRF1, SUVH9 and SUVH2. SRA domain of UHRF1 protein can connect to hydroxymethylated DNA having the similar affinity with methylated one [200]. UHRF1 protein is known to be an epigenetic regulator involved in the association of DNA methylation with chromatin organisation. UHRF1 maintains global and regional methylation pattern of the genome so playing a role for epigenetic inheritance [221, 222] as UHRF1 targeting both DNMT1 and histone lysine methylation forms a link between methylated DNA and histones (Figure 7C) [223-225]. UHRF1 binds to DNMT1 via its SRA domain [226], and even significantly increases DNMT1s activity [227]. Disruption of the complex of UHRF1 with PCNA (proliferating cell nuclear antigen) and DNMT1 was found to stimulate carcinogenesis [228, 229]. However, the overexpressed UHRF1 has been shown to induce global hypomethylation in cancer [230]. This can suggest a dual function of UHRF1 protein similar to some MBD proteins and Kaiso. In *Arabidopsis,* SUVH9 protein prefers to bind to methylated CpGs, but SUVH2 binds asymmetrically patterned of DNA methylation [231].

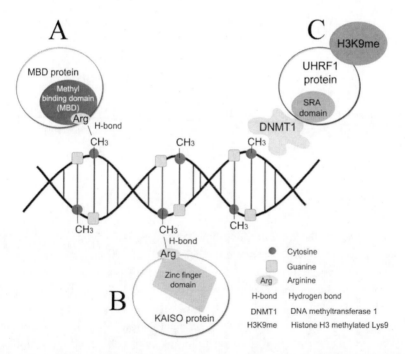

Figure 7. Examples for the binding of methylated DNA-specific proteins to methylated DNA. (A) MBD proteins, (B) Kaiso protein, and (C) UHRF1 protein use different regions; a MBD domain, a zing finger domain and a SRA domain, respectively, to bind methylated DNA. MBD protein and Kaiso proteins bind via such arginine amino acid. UHRF1 differentially binds both DNMT1 and H3K9me.

6.2. Relationship between Methyl-Binding Proteins and DNA Methylation in Different Genomic Regions

Besides the polarity and chemical bonding features, the other factor for determining the binding properties of the protein to modified DNA is the CpG content of the gene/region of interest within the genome. For instance, MBD proteins, with the exception of MBD3, have a higher affinity to the regions with high-content CpGs which are mostly gene promoters [203]. In addition to the targeting methylated CpGs, MBD proteins also regard further specificities. For instance, MeCP2 likes to bind CpGs which have A-T stretch surroundings [232], whereas MBD1 binds to the methylated cytosines in CpGs if these CpGs exist within TG<u>C</u>GCA or T<u>C</u>GCA sequences *(methylated cytosines underlined)* [233]. MBD2, however, selects GGAT<u>C</u>GGCTC sequence of a promoter around methylated CpGs [192]. These can indicate the high specificity of MBD domain for binding to the methylated cytosines regardless with a strict rule that CG pairs should be in a precise repeated context. Similarly, some transcription factors, such PPARG (a protein complex which a nuclear hormone receptor-encoded protein – PPAR – involved), prefer to bind the CpG-free regions between methylated regions [202]. PPAR was also found to involve in 5hmC production from 5meC around its binding site [234]. Zfp57, a zing finger protein functioning in gene repression during early development and maintenance of imprinting, can recognise the methylated CG in TGC<u>C</u>GC sequence through two contacts between ZF2 and ZF3 (flanking zing finger regions of its DNA-binding domain) and 5'TGC and 3'<u>C</u>GC, respectively [220].

6.3. Histones

Histones are the proteins involving in packaging ~ 2 meters-long DNA molecule to fit it into a nucleus. These form chromatin structures as nucleosomes followed by bending nucleosomes. Histone proteins are alkaline proteins, and they can undergo some further modifications including (a) acetylation and (b) methylation of histones. (a) Binding of acetyl group (– COCH$_3$) to histones is associated with gene activation [159], and consequently the removal of these groups (deacetylation) is involved in transcriptional silencing. The effect of acetylation on gene regulation has been proposed to occur by changing the charge from positive to negative via negatively-charged acetyl group [235]. Therefore, the loosen chromatin allows transcription factors to access to DNA. This is interesting because the acetyl group has hydrophilic properties, and histones would also be expected to induce extra polarity due to their acidic characteristics. This may suggest that the electric charges can be more effective than chemical characteristics of compounds on the decision of biological events, if particularly these compounds are highly compact such within chromatin. The other modification, (b) methylation of histones, is involved in both transcriptional activity and repression. Most common methylation occurs at lysine (K) residue of histones, and the function of this methylation varies depending on the number of methyl group added to lysine, the position of lysine and the histone protein type. For instance, one methyl group addition to lysine of histone H4 (H4K20me1) is related to transcription silencing [236]. However, trimethylation of histone H3 (H3K4me3) was found in actively expressed genes [237]. The effect of methylation on histones appears more complex than on DNA since further features are provided by amino acid residues and histone itself, and the strong relationship between histone modifications and cytosine modifications [132, 157, 222, 238, 239] adds more complicated mechanisms to understand epigenetic regulation of genes.

6.4. Transcription Factors

Transcription factors (TFs) involved in active gene expression likely prefer to bind to DNA via H-bond formation [240] rather than hydrophobic interaction so that expected not tend to contact with methylated DNA. However, the context of cytosine modifications in palindromic or chimeric DNA motifs can associate with the binding of transcriptional activator factors. (*Palindromic sequences*: complementary DNA sequences with the same sequence both from 5' to 3' and from 3' to 5' e.g., 5'-AGCT-3' and 3'-TCGA-5,' *chimeric sequences:* complementary DNA sequences that are not in palindromic context).

A remarkable example of transcription activator proteins bound to methylated cytosines is represented by the C/EBP family which involve in differentiation. A new study used *in vitro* oligonucleotides including homotypic or heterotypic CGs (with the same or different combinations of cytosine modifications in both strands, respectively) to examine the binding property of C/EBPβ transcription factor to DNA showed that four homotypic combinations (C/C, 5meC/5meC, 5fC/5fC and 5caC/5caC) in both strands of palindromic motif, TTGC|GCAA, increased C/EBPβ binding, but 5hmC/5hmC decreased [241]. Homotypic strands including 5meC/5meC, 5fC/5fC and 5caC/5caC also induced an increased binding of protein to the chimeric motif (TTGC|GTCA), whereas 5hmC/5hmC again reduced the binding rate [241]. This confirms the previous data with improved C/EBPβ's binding by 5meC [242]. However, 5hmC did not inhibit the binding of another in C/EBP family, C/EBPα, suggesting the selectivity of 5hmC for protein interaction [241]. Heterotypic combinations, 5hmC/C and 5caC/C, enhanced binding to DNA more than C/C [241]. 5meC also improved the other member of C/EBP, C/EBPα's binding [103, 242] in particular to a palindromic motif, TGACGTCA [103]. But it reduced some other TFs' (ATF4, JUN, JUND, CEBPδ, CEBPγ) binding [242]. Methylation of cytosines did induce 10-fold increase in the binding of heterodimer (C/EBPβ|ATF4) to CGAT|GCAA sequence [242]. Similar methodology showed cytosine methylation *(underlined)* in **C**G**C**AG|GTG motif did not alter the binding of a helix-loop-helix heterodimer (Tcf3|Ascl1), whereas methylation in the midpoint of CGCA**C**|GTC motif inhibited the contact between Tcf3|Ascl1 and DNA [243]. Carboxylation of cytosines resulted in a 10-fold increase in binding of Tcf3|Ascl1, and also other homo/heterodimers of Tcf3, Tcf4, Tcf12 and Ascl1 but less than Tcf3|Ascl1 [243]. In addition, methylation has been shown to lessen the binding of TFs involved in housekeeping processes, including ETS, SP1 and NRF-1, to CCGGAAGT, CCCGCC and CGCCTGCG, respectively [244]. These conclude that transcription can be regulated by cytosine modifications depending on DNA sequence and context. These also imply that cytosine modifications can affect the DNA-protein interaction regardless being in a symmetric pattern over two DNA strands.

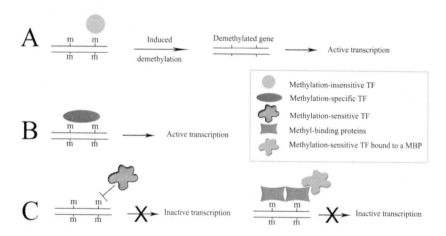

Figure 8. Possible interactions between transcription factors and methylated DNA. Transcription factors (TFs) can be classified as (A) methylation-insensitive TFs, (B) methylation-specific TFs, and (C) methylation-sensitive TFs. These involve in transcriptional regulation. Adapted from Schübeler, 2015 [245].

In general, TFs can be classified within the groups defined by Schübeler regarding their interaction of methylated DNA [245]: (a) methylation-insensitive TFs which bind methylated sites with low CG content, (b) methylation-specific TFs (such C/EBPα and C/EBPβ) which bind methylated cytosines and (c) methylation-sensitive TFs whose binding are blocked either directly by DNA methylation itself (such ATF4, JUND) or indirectly by a methyl-binding protein bound to methylated DNA (Figure 8). A way how TFs can bind to methylated DNA includes that methylation-insensitive TFs facilitate methylation-sensitive TFs' binding through reducing methylation in the site of interest [245] (Figure 8). For instance, *Tat (tyrosine aminotransferase)* gene coding a transcription activator protein is activated by demethylation induced by a methylation-insensitive TF, the glucocorticoid receptor protein [246]. This mechanism is also involved in active demethylation of DNA during DNA repair.

7. BIOLOGICAL SIGNIFICANCE OF CYTOSINE MODIFICATIONS

7.1. In Eukaryotes

The majority of independent reports suggest that the activity of genes can be controlled by the existence or absence of methyl groups (or other functional groups) in CpGs (or other CG context) with regard to positioning throughout the genome [49, 247-250]. For instance, the level of 5hmC in neurons was shown to decrease in 3'-untranslated (UTR) region of *Nr3c1* gene (the gene encoding the glucocorticoid receptor protein) after stress, but its global level did not alter [251]. Although untranslated regions consist of non-coding sequences, they are known to involve in regulation of post-translational events and providing a binding site for regulatory proteins (reviewed in [252]). The polymorphisms in 3'-UTR regions of *i.e., KRAS* [253] and *HLA-G* [254] genes resulted in altered levels of their protein products, and even associated with carcinogenesis [253, 255, 256] indicate the importance of these regions. Besides, the level of CpG methylation was found more stable in exons, introns, CG shores and shelves compared to promoter regions and enhancers [249]. Promoter/enhancer regions are more dynamic regions involving in transcriptional regulation. With the current understandings, we cannot predict the significance of where, when and why methylation occurs in different genomic regions, and not surely know how methylation functions in such gene bodies, but there is some evidence to figure out this. For instance, methylation of CGs in gene bodies, in contrast to being in promoter regions, does not prevent gene expression, even stimulate the transcription [257] consistent with the finding that CpG islands in gene bodies can function as a proxy of promoters [160, 258]. The positive correlation between the increases in methylation at gene bodies and in the gene transcription was reported in human [247, 248] and in *Arabidopsis* [259]. The co-pattern of methylation in both promoter and gene body can be also concluded to be important as patients with schizophrenia showed altered methylation both at promoter regions and gene bodies [250].

Almost one in five autosomal CpGs in DMRs located from distal to transcription sites was detected to be dynamically regulated during normal development of human [249]. In mouse, the amount of allele-specific methylation can depend on the sequence or the pattern of genomic imprinting. In the sequence dependence, the highest methylation occurred at intergenic regions (~45%) followed by introns (38%), transcription start sites (TSS) (8%), transcription end sites (TES) (7.4%) and exons (2%) [260]. On the contrary, depending on parental imprinting, the highest methylation belonged to TSS (~50%) followed by intergenic regions (24%), introns (~12%) and TES (~12%), and the exons (~2%) [260]. Non-CpG methylation (mCHH or mCHG, *H referring A, T or C*) is around 25-36% in mouse neurons [128, 260]. It was revealed that cytosines tend to methylate more in CHH (~25%) than in CHG (~6%) [128, 260], but the highest methylation occurred at CpGs (75%) in neurons [128]. In contrast, imprinted DMRs had the lowest methylation in CpGs both in human (5%) and mouse (4%), but the highest level (~73%) in CHHs in both organisms [260]. These imply that there is no significant contribution of guanine location whether is either next to methylated cytosine or not. Housekeeping genes with a high level of CpGs at their promoters are actively expressed since these CpGs are unmethylated [261]. It can conclude that although the high amount of CpG can indicate a high potential for methylation, it is not always correlated with high methylation. Accordingly, global methylation levels of CpGs and non-CpGs were similar in autosomal chromosomes but different in sex chromosomes [128]. Further, global methylation of non-CpGs was the highest in brain and placenta while CpG methylation was similar in other tissues [128]. Although CpG methylation is persistently high in human neurons, but not in glial cells, from birth to age 50, CH methylation does vary up to 50-years, in particular, with a rapid increase in the first 5 years after birth [262]. Mouse expectedly showed the similar pattern [262]. These can conclude that methylation varies depending on the location of cytosine in the sequence. This also depends on cell or tissue as it is found that only half of CpGs is methylated in frontal cortex, but almost all CpGs are methylated in embryonic fibroblasts [260].

Alterations in methylation pattern have been reported in abnormalities, particularly in neoplasms. Hypermethylation of tumour suppressor gene promoters has been shown to associate with down regulation of genes [34, 263-266]. Decreases in the methylation of proto-oncogene promoters [267-269] or of repetitive DNA elements [25, 270] are also reported in a range of cancers. However, promoter hypermethylation was found not to correlate with silencing of some apoptotic genes in ovarian cancer [271] suggesting that DNA methylation is likely participated in gene regulation depending on the genes and the tissues. DNA methylation profiles can be different between unrelated healthy people (inter-individual differences) [272, 273], and a heterogeneous DNA methylation pattern occurring in healthy cells even from the same organism (intra-individual differences) [273, 274] pointing out that each cell has its unique methylation pattern. It is also remarkable that the variation in methylation pattern is typical even in identical twins sharing the 99% genetic basis [17, 18]. Therefore identical twins represent a good experimental model to reveal the effect of environmental conditions on epigenetic pattern. These variations indicate that interpretation of function(s) and regulations of DNA methylation and other cytosine modifications is not simple.

Apart from 5meC, the pattern of 5hmC has been also reported to alter. As in mammals [45], in zebrafish the level of 5hmC is significantly more (4-fold to 50-fold) in brain tissue than other tissues such liver and testis [275], and the similar pattern of 5hmC in neurons was also seen in amphibians [276]. The level of 5hmC was found low at promoters of such housekeeping genes, whereas pluripotency genes and lineage-specific differentiation related genes had a high level of 5hmC at their promoter regions in nuclear extracts of HeLa cells [52] suggesting the effect of 5hmC on transcription inhibition. Besides, the other modifications, 5fC and 5caC, have been shown to be more enriched in TSSs of active genes than 5meC and 5hmC [277]. 5caC and 5hmC were found to associate with euchromatic regions of the genome in amphibians [278]. 5caC was present at measurable levels in amphibian somatic cells [278], and its level also increased in the cell-specific promoter regions for the duration of neural specification in mouse [74]. Similar to 5caC, the level of 5fC has been shown to increase during neural stem cell specification in mouse, and in repetitive sequences of satellites [76]. While 5caC is the last, but not the least product of the series of reactions beginning with 5meC, the alteration in the level of 5caC can be expected to be inversely correlated with its precursors. Active demethylation can be the good model for both conversions of cytosine modifications to each other in the response of DNA damage and a highly dynamic helix-structure of DNA during active demethylation mechanism (as discussed above). This widely accepted hypothesis for DNA repair includes the erasure of 5meC so that allowing repair factors to access to the damaged site cooperatively driven by TET-mediated catalytic activities and base-excision DNA repair mechanism. The removal of methyl is followed by the replacement with the chemical groups (– OH, – CHO, – COOH, respectively) at the 5th position of cytosine. By the nature of chemical reactions, the level of reactant compound decreases as it is modified to a product molecule. Therefore, in theory, the conversion of 5meC to other modifications should mean the negative correlation between the levels of 5meC and its downstream metabolites. This can be supported by the findings revealed an association of increased 5hmC level with decreased *LINE1* methylation in colorectal cancer [279] suggesting an inverse correlation between those two cytosine modifications. But few cases point out a different tendency. A study showed an insignificant change in the global levels of both 5hmC and 5meC after restraint stress in hippocampus [251] suggesting the possibility that cells can be resistant to alterations in cytosines after DNA damage in tissue-specific manner and/or depending on the damage conditions. Both 5meC and 5hmC levels were low in lung cancerous cells compared to healthy ones [280], and both 5meC and 5hmC levels have been recently shown to be more in DNA samples with oxidative damage obtained from autistic brain of both genders compared to healthy individuals of mice, and the study also included human samples with the similar experimental outcome [281]. Nevertheless, it cannot mean that there is no any conversion of 5meC to 5hmC after DNA damage in these cases, and this is also considerable that in particular although cancer cells present a good model for highly accumulated DNA damage, they are expected to have a depleted DNA repair system. TET enzymes have been evidenced to catalyse the conversion of a cytosine modification to its metabolite(s) [4, 5, 76, 282], and these are accepted as the only responsible enzymes for these reactions as 5meC existed but 5hmC lost in the absence of TET suggesting no conversions that occur without TET [283, 284]. However, parallelism between the level of a precursor molecule and its metabolite can speculate that there are perhaps some alternative ways (by such different enzymes or enzyme variants) to catalyse these molecules.

DNA methylation pattern can vary between the species as the percentage of methylated cytosine found at 14% in *Arabidopsis thaliana*; 7.6% in *Mus musculus*, whereas relatively at very low amount in more primitives as 0.034% in *Drosophila melanogaster* and <0.002% in yeast [285]. This pattern is highly dynamic, especially in human, since not only genotypic factors such ethnicity [286-288], age [289, 290], gender [288], but also environmental factors such diet [291], exercise [292, 293], stress [294], mental issues [295], smoking [269] and even socio-economic status [296] can induce changes in DNA methylation profile at gene specific level and/or in global manner. For instance, methylation of *Alu* region was found to be positively correlated to economic status but *LINE-1* methylation was negatively correlated [287].

7.2. In Prokaryotes

2.3% of cytosines of *Escherichia coli*'s genome are methylated [285]. However, the role of cytosine methylation is quite different in bacteria than eukaryotes. Cytosine methylation functions in restriction modification system of bacteria which protects bacterial DNA from both foreign DNA molecules and restriction enzymes. Nevertheless, 5meC is not the only methylation occurring in the genetic material as differentially from sophisticated organisms, the bacterial genome also consists of 4meC (N4-methylcytosine) and 6mA (N6-methyladenine) modifications playing roles in restriction modification system [297, 298], and therefore maintaining DNA stability. 6mA is also involved in repression of transposon expression, and recently found at high levels in early embryos but less in late embryos of *Drosophila* [299] suggesting its possible role in early development. But 6mA at promoter regions of genes found to be associated with active gene expression [298].

CONCLUSION

The common aspect for gene regulation by DNA methylation is that methyl groups bound to cytosines can recruit a physical barrier between transcription factors and promoter regions of genes and therefore prevents recognition of target DNA sites by transcription factors [300, 301]. However, this should not that simple since methyl groups are not a kind of barrier for binding of some other proteins including MBPs [187, 192, 233, 302], enzymes that are able to remove methylation [80, 303], DNMTs (in particular DNMT1) targeting hemi-methylated DNA [66, 128] and some of transcription factors [103, 241, 242] by their specific characteristics for recognition methylated cytosines.

The whole story sounds the gene regulation by DNA methylation is highly complicated, and importantly there are no strict borders dividing molecular biophysics and (bio)chemistry in terms of defining structure of genetic material. For instance, histone modifications in association with DNA methylation (DNA packaging) are included in both physical and chemical features of genetic material which are significantly involved in transcriptional regulation.

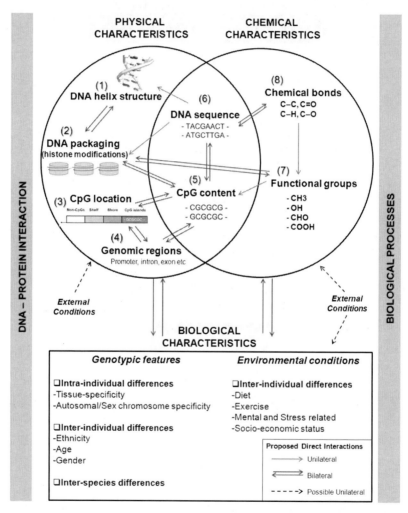

Figure 9. Schematic Representation for Biological Processes Regulated by Cytosine Modifications. This figure shows the physical, chemical (both indicated with numbers) and biological characteristics interacted with cytosine modifications and the proposed interactions (bilateral or unilateral or possible) between and within these characteristics while regulating DNA-protein interaction. This multi-factorial network is concluded to affect biological processes. Some examples for these effects numbered in this figure on biological outcomes are given in *Table 2*.

Table 2. Examples of biological significance by cytosine modifications associated with multiple factors

Physical and/or Chemical Properties	Biolological Characteristics	Significance on Cytosine Modifications
(3) CpG location (4) Genomic region (5) CpG content (7) Functional group	Gene and tissue specificity	Decrease in 5hmC level of 3'-UTR in hippocampus after stress [ref: 251]
	Genomic region specificity	Regulation of human development by 5meC and 5hmC [ref: 249]
	Cell specificity	Association of 5hmC with transcriptional repression [ref: 50]
	Tissue specificity	Alterations in 5meC level associated with schizophrenia [ref: 250]
(4) Genomic region (5) CpG content (6) DNA sequence (7) Functional group	Sequence and allele specificity	Variation in 5meC level as the highest in intergenic, the least in exons (regarding sequence) and the highest in transcription start sites, the least in exons again (regarding genomic imprinting) [ref: 260]
(2) DNA packaging (3) CpG location (4) Genomic regions (5) CpG content (7) Functional group	Inter-species Tissue specificity	Association of histone methylation with CpG density in intra- and intergenic regions [ref: 160]
	Cell specificity	Association of histone acetylation with CpGs in promoters [ref: 159]
(3) CpG location (5) CpG content (7) Functional group	Cell specificity	Difference in 5meC level at CpG and non-CpGs in neuron [refs: 128, 260]
	Tissue specificity	The highest level of 5meC in neuron [ref: 128]
	Cell and tissue specificity	High level 5meC in intra- and intergenic regions [ref: 258]
	Chromosome specificity	Similar level of 5meC in non-CpG and CpG in autosomals, but not in sex chromosomes [ref: 128]
(3) CpG location (4) Genomic region (7) Functional group	Gene specificity	Increased methylation associated with cancer [refs: 28, 34, 263-266]
	Gene specificity	Decreased methylation associated with cancer [refs: 267, 268, 269]
	Genomic region specificity	Decreased methylation associated with cancer [refs: 25, 270]
(4) Genomic region (7) Functional group	Cell specificity	Association of 5hmC in gene body with transcriptional repression [ref: 49]
	Cell specificity	Association of 5fC and 5caC with active genes [ref: 277]
	Gene and cell specificity	Association of 5hmC with gene expression [ref: 183]
(3) CpG location (7) Functional group	Tissue specificity	Difference in 5meC level in human neurons depending on age [ref: 262]
(2) DNA packaging (7) Functional group	Chromosome specificity	Association of histone methylation with 5hmC [ref: 184]
(7) Functional group	Inter-species	Similar high level of 5hmC in mouse [ref: 45], in zebrafish [ref: 275] and amphibians [ref: 276] with in mammals
	Tissue specificity	Decreased levels of 5fC and 5caC in Alzheimer [ref: 48]
	Gene and tissue specificity	Association of 5hmC with gene expression [refs: 45, 46]
	Tissue specificity	Association of 5fC and 5caC with differentiation [ref: 74]
	Inter-individual	Variation in 5meC within unrelateds [refs: 272, 273] and within identical twins [refs: 17, 18]
	Intra-individual	Variation in 5meC level [refs: 273, 274]

In conclusion, the ways that DNA methylation (and other modifications) can affect transcriptional regulation are highly subject to multiple factors as: (1) DNA helix structure, (2) DNA packaging with histones and histone modifications, (3) CpG location, (4) genomic regions, (5) CpG content, (6) DNA sequence, (7) functional groups, and (8) chemical bonds. These factors represent possible pathways for gene regulation by DNA methylation (Figure 9). But these factors, as mentioned before, cannot be thought independent so that they are directly or indirectly interacted with each other, and direct interactions consist of possible bilateral or unilateral interactions which play a role(s) in the mechanism of biological regulation. This is also noted that some external conditions (e.g., stress) can alter these factors and therefore indirectly or directly affect DNA methylation pattern. These determinants in complex relationships result in biological significance through modifying the contact between DNA and proteins (Figure 9), and biological characteristics defined by genotypic and environmental conditions are also important as in the association with physical and chemical effects of cytosine modifications. In this manner, examples of biological significance by cytosine modifications are given in Table 2.

With the exciting completion of Human Genome Project in 2003, all genetic information of human has been available, however there has been still increasing numbers of questions about the human genome. These are based on not the full sequences of DNA strands, but based on (i) modifications of DNA bases, e.g., at cytosine with no sequence change and (ii) a complicated 3D structure of DNA itself, (iii) specific DNA bases/sites with a wide range of proteins such chromatin proteins and transcriptional factors which influence transcriptional regulation. Herein, the perspective of epigenetics presents further information than which provided by genetics i.e., Human Genome Project, and suggests the requirement of "Human Epigenome Project" completed that reveals all information and interactions about cytosine modifications as well as histone modifications. In the future, complete profiles of genome, proteome and such connecting those two, epigenome, will improve the understanding of genetic inheritance and maintenance in complex dynamics controlling the meaning of life.

REFERENCES

[1] Tahiliani M, Koh KP, Shen YH, Pastor WA, Bandukwala H, Brudno Y, et al. Conversion of 5-Methylcytosine to 5-Hydroxymethylcytosine in Mammalian DNA by MLL Partner TET1. *Science* 2009;324:930-5.

[2] Ito S, D'Alessio AC, Taranova OV, Hong K, Sowers LC, Zhang Y. Role of Tet proteins in 5mC to 5hmC conversion, ES-cell self-renewal and inner cell mass specification. *Nature* 2010;466:1129-U151.

[3] Guo JU, Su YJ, Zhong C, Ming GL, Song HJ. Hydroxylation of 5-Methylcytosine by TET1 Promotes Active DNA Demethylation in the Adult Brain. *Cell* 2011;145:423-34.

[4] He YF, Li BZ, Li Z, Liu P, Wang Y, Tang QY, et al. Tet-Mediated Formation of 5-Carboxylcytosine and Its Excision by TDG in Mammalian DNA. *Science* 2011;333:1303-7.

[5] Ito S, Shen L, Dai Q, Wu SC, Collins LB, Swenberg JA, et al. Tet Proteins Can Convert 5-Methylcytosine to 5-Formylcytosine and 5-Carboxylcytosine. *Science* 2011;333:1300-3.

[6] Koh KP, Yabuuchi A, Rao S, Huang Y, Cunniff K, Nardone J, et al. Tet1 and Tet2 Regulate 5-Hydroxymethylcytosine Production and Cell Lineage Specification in Mouse Embryonic Stem Cells. *Cell Stem Cell* 2011;8:200-13.

[7] Maiti A, Drohat AC. Thymine DNA Glycosylase Can Rapidly Excise 5-Formylcytosine and 5-Carboxylcytosine POTENTIAL IMPLICATIONS FOR ACTIVE DEMETHYLATION OF CpG SITES. *J Biol Chem* 2011;286:35334-8.

[8] Pfaffeneder T, Hackner B, Truss M, Munzel M, Mullcr M, Deiml CA, et al. The Discovery of 5-Formylcytosine in Embryonic Stem Cell DNA. *Angew Chem Int Edit* 2011;50:7008-12.

[9] Li E, Bestor TH, Jaenisch R. Targeted Mutation of the DNA Methyltransferase Gene Results in Embryonic Lethality. *Cell* 1992;69:915-26.

[10] Kobayashi H, Sakurai T, Imai M, Takahashi N, Fukuda A, Yayoi O, et al. Contribution of Intragenic DNA Methylation in Mouse Gametic DNA Methylomes to Establish Oocyte-Specific Heritable Marks. *Plos Genet* 2012;8:e1002440.

[11] Chadwick BP, Willard HF. Multiple spatially distinct types of facultative heterochromatin on the human inactive X chromosome. *P Natl Acad Sci USA* 2004;101:17450-5.

[12] Cotton AM, Price EM, Jones MJ, Balaton BP, Kobor MS, Brown CJ. Landscape of DNA methylation on the X chromosome reflects CpG density, functional chromatin state and X-chromosome inactivation. *Hum Mol Genet* 2015;24:1528-39.

[13] Sun BW, Ito M, Mendjan S, Ito Y, Brons IGM, Murrell A, et al. Status of Genomic Imprinting in Epigenetically Distinct Pluripotent Stem Cells. *Stem Cells* 2012;30:161-8.

[14] Buckberry S, Miotto T, Hiendleder S, Roberts C. Quantitative Allele-Specific Expression and DNA Methylation Analysis of H19, IGF2 and IGF2R in the Human Placenta across Gestation Reveals H19 Imprinting Plasticity. *Plos One* 2012;7:e51210.

[15] Ghosh S, Yates AJ, Fruhwald MC, Miecznikowski JC, Plass C, Smiraglia DJ. Tissue specific DNA methylation of CpG islands in normal human adult somatic tissues distinguishes neural from non-neural tissues. *Epigenetics-Us* 2010;5.

[16] Schilling E, Rehli M. Global, comparative analysis of tissue-specific promoter CpG methylation. *Genomics* 2007;90:314-23.

[17] Mill J, Dempster E, Caspi A, Williams B, Moffitt T, Craig I. Evidence for Monozygotic Twin (MZ) Discordance in Methylation Level at Two CpG Sites in the Promoter Region of the Catechol-O-Methyltransferase (COMT) Gene. *American Journal of Medical Genetics Part B (Neuropsychiatric Genetics)* 2006;141B:421-5.

[18] van Dongen J, Ehli EA, Slieker RC, Bartels M, Weber ZM, Davies GE, et al. Epigenetic Variation in Monozygotic Twins: A Genome-Wide Analysis of DNA Methylation in Buccal Cells. *Genes-Basel* 2014;5:347-65.

[19] Schreiner F, El-Maarri O, Gohlke B, Stutte S, Nuesgen N, Mattheisen M, et al. Association of COMT genotypes with S-COMT promoter methylation in growth-discordant monozygotic twins and healthy adults. *Bmc Med Genet* 2011;12.

[20] Zhao JY, Goldberg J, Bremner JD, Vaccarino V. Global DNA Methylation Is Associated With Insulin Resistance A Monozygotic Twin Study. *Diabetes* 2012;61:542-6.

[21] Schneider E, Pliushch G, El Hajj N, Galetzka D, Puhl A, Schorsch M, et al. Spatial, temporal and interindividual epigenetic variation of functionally important DNA methylation patterns. *Nucleic Acids Res* 2010;38:3880-90.

[22] Wagner JR, Busche S, Ge B, Kwan T, Pastinen T, Blanchette M. The relationship between DNA methylation, genetic and expression interindividual variation in untransformed human fibroblasts. *Genome Biology* 2014;15.
[23] Glossop J, Nixon N, Emes R, Haworth K, Packham J, Dawes P, et al. Epigenome-wide profiling identifies significant differences in DNA methylation between matched-pairs of T- and B-lymphocytes from healthy individuals. *Epigenetics* 2013;8:1188-97.
[24] Deaton A, Webb S, Kerr A, Illingworth R, Guy J, Andrews R, et al. Cell type–specific DNA methylation at intragenic CpG islands in the immune system. *Genome Research* 2013;21:1074-86.
[25] Baba Y, Watanabe M, Murata A, Shigaki H, Miyake K, Ishimoto T, et al. LINE-1 Hypomethylation, DNA Copy Number Alterations, and CDK6 Amplification in Esophageal Squamous Cell Carcinoma. *Clin Cancer Res* 2014;20:1114-24.
[26] Wei J, Han B, Mao XY, Wei MJ, Yao F, Jin F. Promoter methylation status and expression of estrogen receptor alpha in familial breast cancer patients. *Tumor Biology* 2012;33:413-20.
[27] Yamaguchi T, Mukai H, Yamashita S, Fujii S, Ushijima T. Comprehensive DNA Methylation and Extensive Mutation Analyses of HER2-Positive Breast Cancer. *Oncology* 2015;88.
[28] Li Z, Lei HZ, Luo M, Wang Y, Dong L, Ma YN, et al. DNA methylation downregulated mir-10b acts as a tumor suppressor in gastric cancer. *Gastric Cancer* 2015;18:43-54.
[29] Bradley-Whiteman M, Lovell M. Epigenetic changes in the progression of Alzheimer's disease. *Mechanisms of Ageing and Development* 2013;134:486-95.
[30] Yang G, Wang J, Shi XY, Yang XF, Ju J, Liu YJ, et al. Detection of global DNA hypomethylation of peripheral blood lymphocytes in patients with infantile spasms. *Epilepsy Res* 2015;109:28-33.
[31] Miller-Delaney SFC, Bryan K, Das S, McKiernan RC, Bray IM, Reynolds JP, et al. Differential DNA methylation profiles of coding and non-coding genes define hippocampal sclerosis in human temporal lobe epilepsy. *Brain* 2014.
[32] Kiss N, Muth A, Andreasson A, Juhlin C, Geli J, Backdahl M, et al. Acquired hypermethylation of the P16INK4A promoter in abdominal paraganglioma: relation to adverse tumor phenotype and predisposing mutation. *Endocrine-Related Cancer* 2013;20:65-78.

[33] Marsit C, Posner M, McClean M, Kelsey K. Hypermethylation of E-Cadherin Is an Independent Predictor of Improved Survival in Head and Neck Squamous Cell Carcinoma. *Cancer* 2008;113:1566-71.

[34] Mir R, Ahmad I, Javid J, Farooq S, Yadav P, Zuberi M, et al. Epigenetic Silencing of DAPK1 Gene is Associated with Faster Disease Progression in India Populations with Chronic Myeloid Leukemia. *Journal of Cancer Science and Therapy* 2013;5:144-9.

[35] Chmelarova M, Krepinska E, Spacek J, Laco J, Beranek M, Palicka V. Methylation in the p53 promoter in epithelial ovarian cancer. *Clin Transl Oncol* 2013;15:160-3.

[36] Pogribny IP, Pogribna M, Christman JK, James SJ. Single-site methylation within the p53 promoter region reduces gene expression in a reporter gene construct: Possible in vivo relevance during tumorigenesis. *Cancer Res* 2000;60:588-94.

[37] Hu SL, Kong XY, Cheng ZD, Sun YB, Shen G, Xu WP, et al. Promoter methylation of p16, Runx3, DAPK and CHFR genes is frequent in gastric carcinoma. *Tumori* 2010;96:726-33.

[38] Zhang J, Yu XL, Zheng GF, Zhao F. DAPK promoter methylation status correlates with tumor metastasis and poor prognosis in patients with non-small cell lung cancer. *Cancer Biomark* 2015;15:609-17.

[39] Luo J, Li YN, Wang F, Zhang WM, Geng X. S-adenosylmethionine inhibits the growth of cancer cells by reversing the hypomethylation status of c-myc and H-ras in human gastric cancer and colon cancer. *Int J Biol Sci* 2010;6:784-95.

[40] Sardi I, DalCanto M, Bartoletti R, Montali E. Abnormal c-myc oncogene DNA methylation in human bladder cancer: Possible role in tumor progression. *Eur Urol* 1997;31:224-30.

[41] Kiran M, Chawla YK, Kaur J. Methylation profiling of tumor suppressor genes and oncogenes in hepatitis virus-related hepatocellular carcinoma in northern India. *Cancer Genet Cytogen* 2009;195:112-9.

[42] Tommasi S, Pinto R, Petriella D, Pilato B, Lacalamita R, Santini D, et al. Oncosuppressor Methylation: A Possible Key Role in Colon Metastatic Progression. *J Cell Physiol* 2011;226:1934-9.

[43] Raynal NJM, Si J, Taby RF, Gharibyan V, Ahmed S, Jelinek J, et al. DNA Methylation Does Not Stably Lock Gene Expression but Instead Serves as a Molecular Mark for Gene Silencing Memory. *Cancer Res* 2012;72:1170-81.

[44] Hon G, Hawkins D, Caballero O, Lo C, Lister R, Pelizzola M, et al. Global DNA hypomethylation coupled to repressive chromatin domain formation and gene silencing in breast cancer. *Genome Research* 2012;22.

[45] Kinney SM, Chin HG, Vaisvila R, Bitinaite J, Zheng Y, Esteve PO, et al. Tissue-specific Distribution and Dynamic Changes of 5-Hydroxymethylcytosine in Mammalian Genomes. *J Biol Chem* 2011;286:24685-93.

[46] Mellen M, Ayata P, Dewell S, Kriaucionis S, Heintz N. MeCP2 Binds to 5hmC Enriched within Active Genes and Accessible Chromatin in the Nervous System. *Cell* 2012;151:1417-30.

[47] Bocker MT, Tuorto F, Raddatz G, Musch T, Yang FC, Xu MJ, et al. Hydroxylation of 5-methylcytosine by TET2 maintains the active state of the mammalian HOXA cluster. *Nat Commun* 2012;3.

[48] Condlife D, Wong A, Troakes C, Proitsi P, Patel Y, Chouliaras L, et al. Cross-region reduction in 5-hydroxymethylcytosine in Alzheimer's disease brain. *Neurobiology of Aging* 2014;35:1850-4.

[49] Robertson J, Robertson AB, Klungland A. The presence of 5-hydroxymethylcytosine at the gene promoter and not in the gene body negatively regulates gene expression. *Biochem Bioph Res Co* 2011;411:40-3.

[50] Williams K, Christensen J, Pedersen MT, Johansen JV, Cloos PAC, Rappsilber J, et al. TET1 and hydroxymethylcytosine in transcription and DNA methylation fidelity. *Nature* 2011;473:343-U472.

[51] Wu H, D'Alessio AC, Ito S, Xia K, Wang ZB, Cui KR, et al. Dual functions of Tet1 in transcriptional regulation in mouse embryonic stem cells. *Nature* 2011;473:389-U578.

[52] Robertson AB, Dahl JA, Vagbo CB, Tripathi P, Krokan HE, Klungland A. A novel method for the efficient and selective identification of 5-hydroxymethylcytosine in genomic DNA. *Nucleic Acids Research* 2011;39.

[53] Newman MR, Sykes PJ, Blyth BJ, Bezak E, Lawrence MD, Morel KL, et al. The Methylation of DNA Repeat Elements is Sex-Dependent and Temporally Different in Response to X Radiation in Radiosensitive and Radioresistant Mouse Strains. *Radiat Res* 2014;181:65-75.

[54] Rai K, Huggins IJ, James SR, Karpf AR, Jones DA, Cairns BR. DNA Demethylation in Zebrafish Involves the Coupling of a Deaminase, a Glycosylase, and Gadd45. *Cell* 2008;135:1201-12.

[55] Simpkins SB, Bocker T, Swisher EM, Mutch DG, Gersell DJ, Kovatich AJ, et al. MLH1 promoter methylation and gene silencing is the primary cause of microsatellite instability in sporadic endometrial cancers. *Hum Mol Genet* 1999;8:661-6.

[56] Wheeler HL, Johnson TB. Researches on Pyrimidines: 5-Methylcytosine. *Am Chem J* 1904;31.

[57] Liutkeviciute Z, Kriukiene E, Licyte J, Rudyte M, Urbanaviciute G, Klimasauskas S. Direct Decarboxylation of 5-Carboxylcytosine by DNA C5-Methyltransferases. *J Am Chem Soc* 2014;136:5884-7.

[58] Mahmood L. The metabolic processes of folic acid and Vitamin B12 deficiency. *Journal of Health Research and Reviews* 2014;1.

[59] Okano M, Bell DW, Haber DA, Li E. DNA methyltransferases Dnmt3a and Dnmt3b are essential for de novo methylation and mammalian development. *Cell* 1999;99:247-57.

[60] Chen T, Ueda Y, Dodge J, Wang Z, Li e. Establishment and maintenance of genomic methylation patterns in mouse embryonic stem cells by Dnmt3a and Dnmt3b. *Mol Cell Biol* 2003;23:5594-605.

[61] Suetake I, Shinozaki F, Miyagawa J, Takeshima H, Tajima S. DNMT3L Stimulates the DNA Methylation Activity of Dnmt3a and Dnmt3b through a Direct Interaction. *The Journal of Biological Chemistry* 2004;279:27816–23.

[62] Xie ZH, Huang YN, Chen ZX, Riggs AD, Ding JP, Gowher H, et al. Mutations in DNA methyltransferase DNMT3B in ICF syndrome affect its regulation by DNMT3L. *Hum Mol Genet* 2006;15:1375-85.

[63] Branco M, Oda M, Reik W. Safeguarding parental identity: Dnmt1maintains imprints during epigenetic reprogramming in early embryogenesis. *Genes Dev* 2008;2008:1567-71.

[64] Robert MF, Morin S, Beaulieu N, Gauthier F, Chute IC, Barsalou A, et al. DNMT1 is required to maintain CpG methylation and aberrant gene silencing in human cancer cells. *Nature Genetics* 2003;33:61-5.

[65] Schermelleh L, Haernmer A, Spada F, Rosing N, Meilinger D, Rothbauer U, et al. Dynamics of Dnmt1 interaction with the replication machinery and its role in postreplicative maintenance of DNA methylation. *Nucleic Acids Res* 2007;35:4301-12.

[66] Feng J, Zhou Y, Campbell SL, Le T, Li E, Sweatt JD, et al. Dnmt1 and Dnmt3a maintain DNA methylation and regulate synaptic function in adult forebrain neurons. *Nature Neuroscience* 2010;13:423-U37.

[67] Thiagarajan D, Dev RR, Khosla S. The DNA methyltranferase Dnmt2 participates in RNA processing during cellular stress. *Epigenetics-Us* 2011;6:103-13.
[68] Kiani J, Grandjean V, Liebers R, Tuorto F, Ghanbarian H, Lyko F, et al. RNA-Mediated Epigenetic Heredity Requires the Cytosine Methyltransferase Dnmt2. *Plos Genet* 2013;9.
[69] Tuorto F, Herbst F, Alerasool N, Bender S, Popp O, Federico G, et al. The tRNA methyltransferase Dnmt2 is required for accurate polypeptide synthesis during haematopoiesis. *Embo J* 2015;34:2350-62.
[70] Seritrakul P, Gross JM. Expression of the De Novo DNA Methyltransferases (dnmt3 - dnmt8) During Zebrafish Lens Development. *Dev Dynam* 2014;243:350-6.
[71] Penn NW, Suwalski R, O'Riley C, Bojanowski K, Yura R. The presence of 5-hydroxymethylcytosine in animal deoxyribonucleic acid. *The Biochemical Journal* 1972;126:781-90.
[72] Barreto G, Schafer A, Marhold J, Stach D, Swaminathan SK, Handa V, et al. Gadd45a promotes epigenetic gene activation by repair-mediated DNA demethylation. *Nature* 2007;445:671-5.
[73] Cortellino S, Xu JF, Sannai M, Moore R, Caretti E, Cigliano A, et al. Thymine DNA Glycosylase Is Essential for Active DNA Demethylation by Linked Deamination-Base Excision Repair. *Cell* 2011;146:67-79.
[74] Wheldon L, Abakir A, Ferjentsik Z, Dudnakova T, Strohbuecker S, D. DC, et al. Transient Accumulation of 5-Carboxylcytosine Indicates Involvement of Active Demethylation in Lineage Specification of Neural Stem Cells. *Cell Reports* 2014;7:1353–61.
[75] Santos F, Peat J, Burgess H, Rada C, Reik W, Dean W. Active demethylation in mouse zygotes involves cytosine deamination and base excision repair. *Epigenetics and Chromatin* 2013;6.
[76] Shen L, Wu H, Diep D, Yamaguchi S, D'Alessio AC, Fung HL, et al. Genome-wide Analysis Reveals TET- and TDG-Dependent 5-Methylcytosine Oxidation Dynamics. *Cell* 2013;153:692-706.
[77] Branco MR, Ficz G, Reik W. Uncovering the role of 5-hydroxymethylcytosine in the epigenome. *Nature Reviews Genetics* 2012;13:7-13.
[78] Kemmerich K, Dingler FA, Rada C, Neuberger MS. Germline ablation of SMUG1 DNA glycosylase causes loss of 5-hydroxymethyluracil- and UNG-backup uracil-excision activities and increases cancer predisposition of Ung−/−Msh2−/− mice. *Nucleic Acids Res* 2012:1-10.

[79] Masaoka A, Matsubara M, Hasegawa R, Tanaka T, Kurisu S, Terato H, et al. Mammalian 5-Formyluracil-DNA glycosylase. 2. Role of SMUG1 uracil-DNA glycosylase in repair of 5-formyluracil and other oxidized and deaminated base lesions. *Biochemistry* 2003;42:5003-12.

[80] Dominguez MD, Teater M, Chambwe N, Redmond D, Vuong B, Chaudhuri J, et al. Demethylase Activity of Aid during Germinal Center B Cell Maturation Could Contribute to Lymphomagenesis. *Blood* 2014;124.

[81] Franchini DM, Chan CF, Morgan H, Incorvaia E, Rangam G, Dean W, et al. Processive DNA Demethylation via DNA Deaminase-Induced Lesion Resolution. *PLoS ONE* 2014;9.

[82] Nabel CS, Jia HJ, Ye Y, Shen L, Goldschmidt HL, Stivers JT, et al. AID/APOBEC deaminases disfavor modified cytosines implicated in DNA demethylation. *Nat Chem Biol* 2012;8:751-8.

[83] Sadakierska-Chudy A, Kostrzewa RM, Filip M. A Comprehensive View of the Epigenetic Landscape Part I: DNA Methylation, Passive and Active DNA Demethylation Pathways and Histone Variants. *Neurotox Res* 2015;27:84-97.

[84] Ehrlich M, Wilson GG, Kuo KC, Gehrke CW. N4-Methylcytosine as a Minor Base in Bacterial-DNA. *J Bacteriol* 1987;169:939-43.

[85] Wolffe A, Jones P, Wade P. DNA Demethylation. *Proc Natl Acad Sci USA* 1999;96:5894-6.

[86] Reis R, Goldstein S. Mitochondrial DNA in Mortal and Immortal Human Cells: genome number, integrity, and methylation. *The Journal of Biological Chemistry* 1983;258:9078-85.

[87] Kelly R, Mahmud A, McKenzie M, Trounce I, St John J. Mitochondrial DNA copy number is regulated in a tissue specific manner by DNA methylation of the nuclear-encoded DNA polymerase gamma A. *Nucleic Acids Res* 2012:1-15.

[88] Ahlert D, Stegemann S, Kahlau S, Ruf S, Bock R. Insensitivity of chloroplast gene expression to DNA methylation. *Mol Genet Genomics* 2009;282:17-24.

[89] Jurkowski TP, Meusburger M, Phalke S, Helm M, Nellen W, Reuter G, et al. Human DNMT2 methylates tRNA(Asp) molecules using a DNA methyltransferase-like catalytic mechanism. *Rna-a Publication of the Rna Society* 2008;14:1663-70.

[90] Schaefer M, Pollex T, Hanna K, Tuorto F, Meusburger M, Helm M, et al. RNA methylation by Dnmt2 protects transfer RNAs against stress-induced cleavage. *Genes & Development* 2010;24:1590-5.

[91] Oswald J, Engemann S, Lane N, Mayer W, Olek A, Fundele R, et al. Active demethylation of the paternal genome in the mouse zygote. *Current Biology* 2000;10:475-8.
[92] Yamazaki T, Yamagata K, Baba T. Time-lapse and retrospective analysis of DNA methylation in mouse preimplantation embryos by live cell imaging. *Developmental Biology* 2007;304:409-19.
[93] Morgan H, Santos F, Green K, Dean W, Reik W. Epigenetic Reprogramming in Mammals. H*um Mol Gene*t 2005;14:47-58.
[94] Silva A, Adenot P, Daniel N, Archilla C, Peynot N, Lucci CM, et al. Dynamics of DNA methylation levels in maternal and paternal rabbit genomes after fertilization. *Epigenetics-Us* 2011;6:987-93.
[95] Smith ZD, Chan MM, Mikkelsen TS, Gu HC, Gnirke A, Regev A, et al. A unique regulatory phase of DNA methylation in the early mammalian embryo. *Nature* 2012;484:339-U74.
[96] Biliya S, Bulla L. Genomic Imprinting: The Influence of Diffrential Methylation in The Two Sexes. *Experimental Biology and Medicine* 2010;235:139-47.
[97] Ariel M, Robinson E, Mccarrey JR, Cedar H. Gamete-Specific Methylation Correlates with Imprinting of the Murine Xist Gene. *Nature Genetics* 1995;9:312-5.
[98] Thorvaldsen JL, Duran KL, Bartolomei MS. Deletion of the H19 differentially methylated domain results in loss of imprinted expression of H19 and Igf2. *Genes & Development* 1998;12:3693-702.
[99] O'Doherty AM, Rutledge CE, Sato S, Thakur A, Lees-Murdock DJ, Hata K, et al. DNA methylation plays an important role in promoter choice and protein production at the mouse Dnmt3L locus. *Developmental Biology* 2011;356:411-20.
[100] Shemer R, Birger Y, Riggs AD, Razin A. Structure of the imprinted mouse Snrpn gene and establishment of its parental-specific methylation pattern. *P Natl Acad Sci USA* 1997;94:10267-72.
[101] Song F, Smith JF, Kimura MT, Morrow AD, Matsuyama T, Nagase H, et al. Association of tissue-specific differentially methylated regions (TDMs) with differential gene expression. *P Natl Acad Sci USA* 2005;102:3336-41.
[102] Bar-Nur O, Russ HA, Efrat S, Benvenisty N. Epigenetic Memory and Preferential Lineage-Specific Differentiation in Induced Pluripotent Stem Cells Derived from Human Pancreatic Islet Beta Cells. *Cell Stem Cell* 2011;9:17-23.

[103] Rishi V, Bhattacharya P, Chatterjee R, Rozenberg J, Zhao JF, Glass K, et al. CpG methylation of half-CRE sequences creates C/EBP alpha binding sites that activate some tissue-specific genes. *P Natl Acad Sci USA* 2010;107:20311-6.

[104] Burgers W, Fuks F, Kouzarides T. DNA Methyltransferases get connected to chromatin. *TRENDS in Genetics* 2002;18:275-7.

[105] Manev H, Dzitoyeva S. Progress in mitochondrial epigenetics. *BioMol Concepts* 2013;4:381-9.

[106] Sun ZY, Terragni J, Borgaro JG, Liu YW, Yu L, Guan SX, et al. High-Resolution Enzymatic Mapping of Genomic 5-Hydroxymethylcytosine in Mouse Embryonic Stem Cells (vol 3, pg 567, 2013). *Cell Reports* 2013;3:968.

[107] Maekawa M, Taniguchi T, Higashi H, Sugimura H, Sugano K, Kanno T. Methylation of Mitochondrial DNA Is Not a Useful Marker for Cancer Detection. *Clin Chem* 2004;50:1480-1.

[108] Shock L, Thakkar P, Peterson E, Moran R, Taylor S. DNA methyltransferase 1, cytosine methylation, and cytosine hydroxymethylation in mammalian mitochondria. *PNAS* 2011;108:3630-5.

[109] Pirola C, Gianotti T, Burgueno A, Funes M, Loidl C, Mallardi P, et al. Epigenetic modification of liver mitochondrial DNA is associated with histological severity of nonalcoholic fatty liver disease. *Gut* 2012.

[110] Dimmock D, Tang LY, Schmitt ES, Wong LJC. Quantitative Evaluation of the Mitochondrial DNA Depletion Syndrome. *Clin Chem* 2010;56:1119-27.

[111] Renis M, Cantatore P, Polosa PL, Fracasso F, Gadaleta MN. Content of Mitochondrial-DNA and of 3 Mitochondrial Rnas in Developing and Adult-Rat Cerebellum. *J Neurochem* 1989;52:750-4.

[112] Robin ED, Wong R. Mitochondrial-DNA Molecules and Virtual Number of Mitochondria Per Cell in Mammalian-Cells. *J Cell Physiol* 1988;136:507-13.

[113] Bellizzi D, D'Aquila P, Scafone T, Giordano M, Riso V, Riccio A, et al. The Control Region of Mitochondrial DNA Shows an Unusual CpG and Non-CpG Methylation Pattern. *DNA Res* 2013;20:537-47.

[114] Dzitoyeva S, Chen H, Manev H. Effect of aging on 5-hydroxymethylcytosine in brain mitochondria. *Neurobiol Aging* 2012;33:2881-91.

[115] Wang S, Lv J, Zhang LL, Dou JZ, Sun Y, Li X, et al. MethylRAD: a simple and scalable method for genome-wide DNA methylation profiling using methylation-dependent restriction enzymes. *Open Biol* 2015;5.

[116] Kobayashi H, Ngernprasirtsiri J, Akazawa T. Transcriptional Regulation and DNA Methylation in Plastids during Transitional Conversion of Chloroplasts to Chromoplasts. *Embo J* 1990;9:307-13.

[117] Tuorto F, Liebers R, Musch T, Schaefer M, Hofmann S, Kellner S, et al. RNA cytosine methylation by Dnmt2 and NSun2 promotes tRNA stability and protein synthesis. *Nature Structural & Molecular Biology* 2012;19:900-5.

[118] Matzke MA, Mosher RA. RNA-directed DNA methylation: an epigenetic pathway of increasing complexity. *Nature Reviews Genetics* 2014;15:394-408.

[119] Fu LJ, Guerrero CR, Zhong N, Amato NJ, Liu YH, Liu S, et al. Tet-Mediated Formation of 5-Hydroxymethylcytosine in RNA. *J Am Chem Soc* 2014;136:11582-5.

[120] Zhang HY, Xiong J, Qi BL, Feng YQ, Yuan BF. The existence of 5-hydroxymethylcytosine and 5-formylcytosine in both DNA and RNA in mammals. *Chem Commun* 2016;52:737-40.

[121] Huber SM, van Delft P, Mendil L, Bachman M, Smollett K, Werner F, et al. Formation and Abundance of 5-Hydroxymethylcytosine in RNA. *Chembiochem* 2015;16:752-5.

[122] Chodavarapu RK, Feng SH, Bernatavichute YV, Chen PY, Stroud H, Yu YC, et al. Relationship between nucleosome positioning and DNA methylation. *Nature* 2010;466:388-92.

[123] Carninci P, Sandelin A, Lenhard B, Katayama S, Shimokawa K, Ponjavic J, et al. Genome-wide analysis of mammalian promoter architecture and evolution (vol 38, pg 626, 2006). *Nat Genet* 2007;39:1174-.

[124] Eckhardt F, Lewin J, Cortese R, Rakyan VK, Attwood J, Burger M, et al. DNA methylation profiling of human chromosomes 6, 20 and 22. *Nat Genet* 2006;38:1378-85.

[125] Weber M, Hellmann I, Stadler MB, Ramos L, Paabo S, Rebhan M, et al. Distribution, silencing potential and evolutionary impact of promoter DNA methylation in the human genome. *Nat Genet* 2007;39:457-66.

[126] Saxonov S, Berg P, Brutlag DL. A genome-wide analysis of CpG dinucleotides in the human genome distinguishes two distinct classes of promoters. *P Natl Acad Sci USA* 2006;103:1412-7.

[127] Lokk K, Modhukur V, Rajashekar B, Martens K, Magi R, Kolde R, et al. DNA methylome profiling of human tissues identifies global and tissue-specific methylation patterns. *Genome Biology* 2014;15.

[128] Guo JU, Su YJ, Shin JH, Shin JH, Li HD, Xie B, et al. Distribution, recognition and regulation of non-CpG methylation in the adult mammalian brain. *Nat Neurosci* 2014;17:215-22.

[129] Guo W, Chung W, Qian M, Pellegrini M, Zhang M. Characterizing the strand-specific distribution of non-CpG methylation in human pluripotent cells. *Nucleic Acids Research* 2013.

[130] Fenouil R, Cauchy P, Koch F, Descostes N, Cabeza JZ, Innocenti C, et al. CpG islands and GC content dictate nucleosome depletion in a transcription-independent manner at mammalian promoters. *Genome Research* 2012;22:2399-408.

[131] Yang XJ, Han H, De Carvalho DD, Lay FD, Jones PA, Liang GN. Gene Body Methylation Can Alter Gene Expression and Is a Therapeutic Target in Cancer. *Cancer Cell* 2014;26:577-90.

[132] Hahn MA, Wu XW, Li AX, Hahn T, Pfeifer GP. Relationship between Gene Body DNA Methylation and Intragenic H3K9me3 and H3K36me3 Chromatin Marks. *PLoS ONE* 2011;6.

[133] Waalwijk C, Flavell RA. DNA Methylation at a Ccgg Sequence in Large Intron of Rabbit Beta-Globin Gene - Tissue-Specific Variations. *Nucleic Acids Research* 1978;5:4631-41.

[134] Gelfman S, Cohen N, Yearim A, Ast G. DNA-methylation effect on cotranscriptional splicing is dependent on GC architecture of the exon-intron structure. *Genome Research* 2013;23:789-99.

[135] Amit M, Donyo M, Hollander D, Goren A, Kim E, Gelfman S, et al. Differential GC Content between Exons and Introns Establishes Distinct Strategies of Splice-Site Recognition. *Cell Reports* 2012;1:543-56.

[136] Rademacher K, Schroder C, Kanber D, Klein-Hitpass L, Wallner S, Zeschnigk M, et al. Evolutionary Origin and Methylation Status of Human Intronic CpG Islands that Are Not Present in Mouse. *Genome Biol Evol* 2014;6:1579-88.

[137] Shenker N, Flanagan JM. Intragenic DNA methylation: implications of this epigenetic mechanism for cancer research. *British Journal of Cancer* 2012;106:248-53.

[138] Kulis M, Queiros AC, Beekman R, Martin-Subero JI. Intragenic DNA methylation in transcriptional regulation, normal differentiation and cancer. *Bba-Gene Regul Mech* 2013;1829:1161-74.

[139] Watson JD, Crick FHC. Molecular Structure of Nucleic Acids - a Structure for Deoxyribose Nucleic Acid. *Nature* 1953;171:737-8.
[140] Berg JM, Tymoczko JL, Stryer L. Biochemistry. 5th Edition. New York: *W. H. Freeman and Company*; 2002.
[141] Rohs R, Jin XS, West SM, Joshi R, Honig B, Mann RS. Origins of Specificity in Protein-DNA Recognition. *Annu Rev Biochem* 2010;79:233-69.
[142] Bewley CA, Gronenborn AM, Clore GM. Minor groove-binding architectural proteins: Structure, function, and DNA recognition. *Annu Rev Bioph Biom* 1998;27:105-31.
[143] Gajiwala KS, Chen H, Cornille F, Roques BP, Reith W, Mach B, et al. Structure of the winged-helix protein hRFX1 reveals a new mode of DNA binding. *Nature* 2000;403:916-21.
[144] Sabogal A, Lyubimov AY, Corn JE, Berger JM, Rio DC. THAP proteins target specific DNA sites through bipartite recognition of adjacent major and minor grooves. *Nature Structural & Molecular Biology* 2010;17:117-U45.
[145] Badis G, Berger MF, Philippakis AA, Talukder S, Gehrke AR, Jaeger SA, et al. Diversity and Complexity in DNA Recognition by Transcription Factors. *Science* 2009;324:1720-3.
[146] Afek A, Schipper JL, Horton J, Gordan R, Lukatsky DB. Protein-DNA binding in the absence of specific base-pair recognition. *P Natl Acad Sci USA* 2014;111:17140-5.
[147] Wingender E. Criteria for an Updated Classification of Human Transcription Factor DNA-Binding Domains. *J Bioinf Comput Biol* 2013;11.
[148] Buck-Koehntop BA, Stanfield RL, Ekiert DC, Martinez-Yamout MA, Dyson HJ, Wilson IA, et al. Molecular basis for recognition of methylated and specific DNA sequences by the zinc finger protein Kaiso. *P Natl Acad Sci USA* 2012;109:15229-34.
[149] Rohs R, West SM, Sosinsky A, Liu P, Mann RS, Honig B. The role of DNA shape in protein-DNA recognition. *Nature* 2009;461.
[150] Vargason JM, Eichman BF, Ho PS. The extended and eccentric E-DNA structure induced by cytosine methylation or bromination. *Nature Structural Biology* 2000;7:758-61.
[151] Raiber EA, Murat P, Chirgadze DY, Beraldi D, Luisi BF, Balasubramanian S. 5-Formylcytosine alters the structure of the DNA double helix. *Nature Structural & Molecular Biology* 2015;22:44-9.

[152] Thalhammer A, Hansen AS, El-Sagheer AH, Brown T, Schofield CJ. Hydroxylation of methylated CpG dinucleotides reverses stabilisation of DNA duplexes by cytosine 5-methylation. *Chem Commun* 2011;47:5325-7.

[153] Hahn M, Heinemann U. DNA Helix Structure and Refinement Algorithm - Comparison of Models for D(Ccaggcm(5)Ctgg) Derived from Nuclsq, Tnt and X-Plor. *Acta Crystallogr D* 1993;49:468-77.

[154] Goodsell DS, Kopka ML, Cascio D, Dickerson RE. Crystal-Structure of Catggccatg and Its Implications for a-Tract Bending Models. *P Natl Acad Sci USA* 1993;90:2930-4.

[155] Grzeskowiak K, Goodsell DS, Kaczorgrzeskowiak M, Cascio D, Dickerson RE. Crystallographic Analysis of C-C-a-a-G-C-T-T-G-G and Its Implications for Bending in B-DNA. *Biochemistry-Us* 1993;32:8923-31.

[156] Dickerson RE, Goodsell DS, Neidle S. ... The Tyranny of the Lattice ... *P Natl Acad Sci USA* 1994;91:3579-83.

[157] Jimenez-Useche I, Ke JY, Tian YQ, Shim D, Howell SC, Qiu XY, et al. DNA Methylation Regulated Nucleosome Dynamics. *Scientific Reports* 2013;3.

[158] Portella G, Battistini F, Orozco M. Understanding the Connection between Epigenetic DNA Methylation and Nucleosome Positioning from Computer Simulations. *PLoS Computational Biology* 2013;9.

[159] Karmodiya K, Krebs AR, Oulad-Abdelghani M, Kimura H, Tora L. H3K9 and H3K14 acetylation co-occur at many gene regulatory elements, while H3K14ac marks a subset of inactive inducible promoters in mouse embryonic stem cells. *BMC Genomics* 2012;13.

[160] Illingworth RS, Gruenewald-Schneider U, Webb S, Kerr ARW, James KD, Turner DJ, et al. Orphan CpG Islands Identify Numerous Conserved Promoters in the Mammalian Genome. *Plos Genetics* 2010;6.

[161] Severin PMD, Zou XQ, Gaub HE, Schulten K. Cytosine methylation alters DNA mechanical properties. *Nucleic Acids Research* 2011;39:8740-51.

[162] Mirsaidov U, Timp W, Zou X, Dimitrov V, Schulten K, Feinberg AP, et al. Nanoelectromechanics of Methylated DNA in a Synthetic Nanopore. *Biophys J* 2009;96:L32-L4.

[163] Severin PMD, Zou XQ, Schulten K, Gaub HE. Effects of Cytosine Hydroxymethylation on DNA Strand Separation. *Biophys J* 2013;104:208-15.

[164] Luger K, Mader AW, Richmond RK, Sargent DF, Richmond TJ. Crystal structure of the nucleosome core particle at 2.8 angstrom resolution. *Nature* 1997;389:251-60.

[165] Mondal M, Choudhury D, Chakrabarti J, Bhattacharyya D. Role of indirect readout mechanism in TATA box binding protein–DNA interaction. *Journal of Computer-Aided Molecular Design* 2015;29:283-95.

[166] Slutsky M, Mirny LA. Kinetics of protein-DNA interaction: Facilitated target location in sequence-dependent potential. *Biophys J* 2004;87:4021-35.

[167] Blair RH, Goodrich JA, Kugel JF. Single-Molecule Fluorescence Resonance Energy Transfer Shows Uniformity in TATA Binding Protein-Induced DNA Bending and Heterogeneity in Bending Kinetics. *Biochemistry-Us* 2012;51:7444-55.

[168] Moore JW, Stanitski CL, Jurs PC. Principles of Chemistry: The Molecular Science. 1st ed: Cengage Learning; 2009.

[169] Hart H, Craine LE, Hart DJ, Hadad CM. Organic Chemistry: A Short Course. 12th ed. USA: Brooks/Cole, Cengage Learning; 2006.

[170] Chang R. General Chemistry. 4th ed: MC Graw Hill; 2006.

[171] Schiesser S, Pfaffeneder T, Sadeghian K, Hackner B, Steigenberger B, Schroder AS, et al. Deamination, Oxidation, and C-C Bond Cleavage Reactivity of 5-Hydroxymethylcytosine, 5-Formylcytosine, and 5-Carboxycytosine. *J Am Chem Soc* 2013;135:14593-9.

[172] Lazarovici A, Zhou TY, Shafer A, Machado ACD, Riley TR, Sandstrom R, et al. Probing DNA shape and methylation state on a genomic scale with DNase I. *P Natl Acad Sci USA* 2013;110:6376-81.

[173] Nelson M, Mcclelland M. Site-Specific Methylation - Effect on DNA Modification Methyltransferases and Restriction Endonucleases. *Nucleic Acids Research* 1991;19:2045-75.

[174] Loenen WAM, Dryden DTF, Raleigh EA, Wilson GG, Murray NE. Highlights of the DNA cutters: a short history of the restriction enzymes. *Nucleic Acids Research* 2014;42:3-19.

[175] Sistla S, Rao DN. S-Adenosyl-L-methionine–Dependent Restriction Enzymes. *Critical Reviews in Biochemistry and Molecular Biology* 2004;39:1-19.

[176] Biot C, Wintjens R, Rooman M. Stair motifs at protein-DNA interfaces: Nonadditivity of H-bond, stacking, and cation-pi interactions. *J Am Chem Soc* 2004;126:6220-1.

[177] Rooman M, Lievin J, Buisine E, Wintjens R. Cation-pi/H-bond stair motifs at protein-DNA interfaces. *J Mol Biol* 2002;319:67-76.
[178] Tsai CL, Tainer JA. Probing DNA by 2-OG-Dependent Dioxygenase. *Cell* 2013;155:1448-50.
[179] Umezawa Y, Nishio M. Thymine-methyl/π interaction implicated in the sequence-dependent deformability of DNA. *Nucleic Acids Research* 2002;30:2183-92.
[180] Alabugin IV, Gilmore KM, Peterson PW. Hyperconjugation. *Wires Comput Mol Sci* 2011;1:109-41.
[181] Wongtawan T, Taylor JE, Lawson KA, Wilmut I, Pennings S. Histone H4K20me3 and HP1 alpha are late heterochromatin markers in development, but present in undifferentiated embryonic stem cells. *J Cell Sci* 2011;124:1878-90.
[182] Hajkova P, Ancelin K, Waldmann T, Lacoste N, Lange U, Cesari F, et al. Chromatin Dynamics during Epigenetic Reprogramming in the Mouse Germ Line. *Nature* 2008;452:877-81.
[183] Ficz G, Branco M, Seisenberger S, Santos F, Krueger F, Hore T, et al. Dynamic regulation of 5-hydroxymethylcytosine in mouse ES cells and during differentiation. *Nature* 2011;473:398-402.
[184] Kubiura M, Okano M, Kimura H, Kawamura F, Tada M. Chromosome-wide regulation of euchromatin-specific 5mC to 5hmC conversion in mouse ES cells and female human somatic cells. *Chromosome Research* 2012;20:837-48.
[185] Fournier A, Sasai N, Nakao M, Defossez PA. The role of methyl-binding proteins in chromatin organization and epigenome maintenance. *Brifings in Functional Genomics* 2011.
[186] Sasai N, Nakao M, Defossez PA. Sequence-specific recognition of methylated DNA by human zinc-finger proteins. *Nucleic Acids Research* 2010;38:5015-22.
[187] Zou XQ, Ma W, Solov'yov IA, Chipot C, Schulten K. Recognition of methylated DNA through methyl-CpG binding domain proteins. *Nucleic Acids Research* 2012;40:2747-58.
[188] Badran A, Furman J, Ma A, Comi T, Porter J, Ghosh I. Evaluating the Global CpG Methylation Status of Native DNA Utilizing a Bipartite Split-Luciferase Sensor. *Anal Chem* 2011;83:7151-7.
[189] Yildirim O, Li RW, Hung JH, Chen PB, Dong XJ, Ee LS, et al. Mbd3/NURD Complex Regulates Expression of 5-Hydroxymethylcytosine Marked Genes in Embryonic Stem Cells. *Cell* 2011;147:1498-510.

[190] Laget S, Joulie M, Le Masson F, Sasai N, Christians E, Pradhan S, et al. The Human Proteins MBD5 and MBD6 Associate with Heterochromatin but They Do Not Bind Methylated DNA. *Plos One* 2010;5.

[191] Ohki I, Shimotake N, Fujita N, Jee JG, Ikegami T, Nakao M, et al. Solution structure of the methyl-CpG binding domain of human MBD1 in complex with methylated DNA. *Cell* 2001;105:487-97.

[192] Scarsdale JN, Webb HD, Ginder GD, Williams DCJ. Solution structure and dynamic analysis of chicken MBD2 methyl binding domain bound to a target-methylated DNA sequence. *Nucleic Acids Research* 2011;39:6741-52.

[193] Ho KL, McNae IW, Schmiedeberg L, Klose RJ, Bird AP, Walkinshaw MD. MeCP2 binding to DNA depends upon hydration at methyl-CpG. *Molecular Cell* 2008;29:525-31.

[194] Arakawa T, Tsurnoto K, Nagase K, Ejima D. The effects of arginine on protein binding and elution in hydrophobic interaction and ion-exchange chromatography. *Protein Expres Purif* 2007;54:110-6.

[195] Fauchere JL, Pliska V. Hydrophobic Parameters-Pi of Amino-Acid Side-Chains from the Partitioning of N-Acetyl-Amino-Acid Amides. *Eur J Med Chem* 1983;18:369-75.

[196] Minkovsky A, Sahakyan A, Rankin-Gee E, Bonora G, Patel S, Plath K. The Mbd1-Atf7ip-Setdb1 pathway contributes to the maintenance of X chromosome inactivation. *Epigenetics and Chromatin* 2014;7.

[197] Fujita N, Shimotake N, Ohki I, Chiba T, Saya H, Shirakawa M, et al. Mechanism of Transcriptional Regulation by Methyl-CpG Binding Protein MBD1. *Mol Cell Biol* 2000;20:5107-18.

[198] Ichimura T, Watanabe S, Sakamoto Y, Aoto T, Fujita N, Nakao M. Transcriptional repression and heterochromatin formation by MBD1 and MCAF/AM family proteins. *J Biol Chem* 2005;280:13928-35.

[199] Fujita N, Takebayashi S, Okumura K, Kudo S, Chiba T, Saya H, et al. Methylation-Mediated Transcriptional Silencing in Euchromatin by Methyl-CpG Binding Protein MBD1 isoforms. *Mol Cell Biol* 1999;19:6415-26.

[200] Frauer C, Hoffmann T, Bultmann S, Casa V, Cardoso MC, Antes I, et al. Recognition of 5-Hydroxymethylcytosine by the Uhrf1 SRA Domain. *PLoS ONE* 2011;6.

[201] Khrapunov S, Warren C, Cheng H, Berko ER, Greally JM, Brenowitz M. Unusual characteristics of the DNA binding domain of epigenetic regulatory protein MeCP2 determine its binding specificity. *Biochemistry-Us* 2014;53:3379−91.

[202] Blattler A, Farnham PJ. Cross-talk between Site-specific Transcription Factors and DNA Methylation States. *The Journal of Biological Chemistry* 2013;288:34287-94.

[203] Baubec T, Ivanek R, Lienert F, Schubeler D. Methylation-Dependent and -Independent Genomic Targeting Principles of the MBD Protein Family. *Cell* 2013;153:480-92.

[204] Çelik S, Li Y, O'Neill C. The exit of mouse embryonic fibroblasts from the cell-cycle changes the nature of solvent exposure of the 5´-methylcytosine epitope within chromatin. *PloS ONE* 2014;9 (4):e92523.

[205] Jorgensen HF, Ben-Porath I, Bird AP. Mbd1 is recruited to both methylated and nonmethylated CpGs via distinct DNA binding domains. *Mol Cell Biol* 2004;24:3387-95.

[206] Günther K, Rust M, Leers J, Boettger T, Scharfe M, Jarek M, et al. Differential roles for MBD2 and MBD3 at methylated CpG islands, active promoters and binding to exon sequences. *Nucleic Acids Research* 2013;41:3010-21.

[207] Prokhortchouk AV, Aitkhozhina DS, Sablina AA, Ruzov AS, Prokhortchouk EB. Kaiso, a new protein of the BTB/POZ family, specifically binds to methylated DNA sequences. *Russian Journal of Genetics* 2001;37:603-9.

[208] Liu YW, Zhang X, Blumenthal RM, Cheng XD. A common mode of recognition for methylated CpG. *Trends Biochem Sci* 2013;38:177-83.

[209] Blattler A, Yao L, Wang Y, Ye Z, Jin VX, Farnham PJ. ZBTB33 binds unmethylated regions of the genome associated with actively expressed genes. *Epigenetics and Chromatin* 2013;6.

[210] Lopes EC, Valls E, Figueroa ME, Mazur A, Meng FG, Chiosis G, et al. Kaiso contributes to DNA methylation-dependent silencing of tumor suppressor genes in colon cancer cell lines. *Cancer Research* 2008;68:7258-63.

[211] Koh DI, Han DY, Ryu H, Choi WI, Jeon BN, Kim MK, et al. KAISO, a critical regulator of p53-mediated transcription of CDKN1A and apoptotic genes. *P Natl Acad Sci USA* 2014;111:15078-83.

[212] Hu S, Wan J, Su Y, Song Q, Zeng Y, Nguyen HN, et al. DNA methylation presents distinct binding sites for human transcription factors. *eLife* 2013;3:e00726.

[213] Liu YW, Olanrewaju YO, Zheng Y, Hashimoto H, Blumenthal RM, Zhang X, et al. Structural basis for Klf4 recognition of methylated DNA. *Nucleic Acids Research* 2014;42:4859-67.

[214] Schuetz A, Nana D, Rose C, Zocher G, Milanovic M, Koenigsmann J, et al. The structure of the Klf4 DNA-binding domain links to self-renewal and macrophage differentiation. *Cellular and Molecular Life Sciences* 2011;68:3121-31.
[215] Takahashi K, Yamanaka S. Induction of pluripotent stem cells from mouse embryonic and adult fibroblast cultures by defined factors. *Cell* 2006;126:663-76.
[216] El-Karim EA, Hagos EG, Ghaleb AM, Yu B, Yang VW. Kruppel-like factor 4 regulates genetic stability in mouse embryonic fibroblasts. *Mol Cancer* 2013;12.
[217] Shum CKY, Lau ST, Tsoi LLS, Chan LK, Yam JWP, Ohira M, et al. Kruppel-like factor 4 (KLF4) suppresses neuroblastoma cell growth and determines non-tumorigenic lineage differentiation. *Oncogene* 2013;32:4086-99.
[218] Zammarchi F, Morelli M, Menicagli M, Di Cristofano C, Zavaglia K, Paolucci A, et al. KLF4 is a Novel Candidate Tumor Suppressor Gene in Pancreatic Ductal Carcinoma. *Am J Pathol* 2011;178:361-72.
[219] Dubois-Chevalier J, Oger F, Dehondt H, Firmin FF, Gheeraert C, Staels B, et al. A dynamic CTCF chromatin binding landscape promotes DNA hydroxymethylation and transcriptional induction of adipocyte differentiation. *Nucleic Acids Research* 2014;42:10943-59.
[220] Liu YW, Toh H, Sasaki H, Zhang X, Cheng XD. An atomic model of Zfp57 recognition of CpG methylation within a specific DNA sequence. *Genes & Development* 2012;26:2374-9.
[221] Sharif J, Muto M, Takebayashi SI, Suetake I, Iwamatsu A, Endo TA, et al. The SRA protein Np95 mediates epigenetic inheritance by recruiting Dnmt1 to methylated DNA. *Nature* 2007;450:908-U25.
[222] Liu X, Gao Q, Li P, Zhao Q, Zhang J, Li J, et al. UHRF1 targets DNMT1 for DNA methylation through cooperative binding of hemi-methylated DNA and methylated H3K9. *Nat Commun* 2013;4.
[223] Rose NR, Klose RJ. Understanding the relationship between DNA methylation and histone lysine methylation. *Bba-Gene Regul Mech* 2014;1839:1362-72.
[224] Liu XL, Gao QQ, Li PS, Zhao Q, Zhang JQ, Li JW, et al. UHRF1 targets DNMT1 for DNA methylation through cooperative binding of hemi-methylated DNA and methylated H3K9. *Nat Commun* 2013;4.
[225] Rothbart SB, Krajewski K, Nady N, Tempel W, Xue S, Badeaux AI, et al. Association of UHRF1 with methylated H3K9 directs the maintenance of DNA methylation. *Nat Struct Mol Biol* 2012;19:1155-+.

[226] Berkyurek AC, Suetake I, Arita K, Takeshita K, Nakagawa A, Shirakawa M, et al. The DNA Methyltransferase Dnmt1 Directly Interacts with the SET and RING Finger-associated (SRA) Domain of the Multifunctional Protein Uhrf1 to Facilitate Accession of the Catalytic Center to Hemi-methylated DNA. *J Biol Chem* 2014;289:379-86.

[227] Bashtrykov P, Jankevicius G, Jurkowska RZ, Ragozin S, Jeltsch A. The UHRF1 Protein Stimulates the Activity and Specificity of the Maintenance DNA Methyltransferase DNMT1 by an Allosteric Mechanism. *J Biol Chem* 2014;289:4106-15.

[228] Hervouet E, Lalier L, Debien E, Cheray M, Geairon A, Rogniaux H, et al. Tumor induction by disruption of the Dnmt1, PCNA and UHRF1 interactions. *Nature Preceedings* 2008.

[229] Pacaud R, Brocard E, Lalier L, Hervout E, Vallette FM, Cartron PF. The DNMT1/PCNA/UHRF1 disruption induces tumorigenesis characterised by similar genetic and epigenetic signatures. *Nature* 2014.

[230] Mudbhary R, Hoshida Y, Chernyavskaya Y, Jacob V, Villanueva A, Fie MI, et al. UHRF1 Overexpression Drives DNA Hypomethylation and Hepatocellular Carcinoma. *Cancer Cell* 2014;25:196-209.

[231] Johnson LM, Law JA, Khattar A, Henderson IR, Jacobsen SE. SRA-Domain Proteins Required for DRM2-Mediated De Novo DNA Methylation. *Plos Genet* 2008;4.

[232] Klose RJ, Sarraf SA, Schmiedeberg L, McDermott SM, Stancheva I, Bird AP. DNA binding selectivity of MeCP2 due to a requirement for A/T sequences adjacent to methyl-CpG. *Molecular Cell* 2005;19:667-78.

[233] Clouaire T, Heras JIDL, Merusi C, Stancheva I. Recruitment of MBD1 to target genes requires sequence-specific interaction of the MBD domain with methylated DNA. *Nucleic Acids Research* 2010;38:4620-34.

[234] Fujiki K, Shinoda A, Kano F, Sato R, Shirahige K, Murata M. PPAR gamma-induced PARylation promotes local DNA demethylation by production of 5-hydroxymethylcytosine. *Nature Communications* 2013;4.

[235] Graff J, Tsai LH. Histone acetylation: molecular mnemonics on the chromatin. *Nat Rev Neurosci* 2013;14:97-111.

[236] Kalakonda N, Fischle W, Boccuni P, Gurvich N, Hoya-Arias R, Zhao X, et al. Histone H4 lysine 20 monomethylation promotes transcriptional repression by L3MBTL1. *Oncogene* 2008;27:4293-304.

[237] Santos-Rosa H, Schneider R, Bannister AJ, Sherriff J, Bernstein BE, Emre NCT, et al. Active genes are tri-methylated at K4 of histone H3. *Nature* 2002;419:407-11.

[238] Xu WH, Li ZC, Yu B, He XY, Shi JS, Zhou R, et al. Effects of DNMT1 and HDAC Inhibitors on Gene-Specific Methylation Reprogramming during Porcine Somatic Cell Nuclear Transfer. *PLoS ONE* 2013;8.

[239] Etchegaray JP, Chavez L, Huang Y, Ross KN, Choi J, Martinez-Pastor B, et al. The histone deacetylase SIRT6 controls embryonic stem cell fate via TET-mediated production of 5-hydroxymethylcytosine. *Nature Cell Biology* 2015;17:545-U50.

[240] Liu LA, Bader JS. Ab initio prediction of transcription factor binding sites. *Pacific Symposium on Biocomputing* 2007:484-95.

[241] Sayeed SK, Zhao J, Sathyanarayana BK, Golla JP, Vinson C. C/EBPβ (CEBPB) protein binding to the C/EBP|CRE DNA 8-mer TTGC|GTCA 2 is inhibited by 5hmC and enhanced by 5mC, 5fC, and 5caC in the 3 CG dinucleotide. *Biochimica et Biophysica Acta* 2015;1849(6):583-9.

[242] Mann IK, Chatterjee R, Zhao JF, He XM, Weirauch MT, Hughes TR, et al. CG methylated microarrays identify a novel methylated sequence bound by the CEBPB vertical bar ATF4 heterodimer that is active in vivo. *Genome Research* 2013;23:988-97.

[243] Golla JP, Zhao JF, Mann IK, Sayeed SK, Mandal A, Rose RB, et al. Carboxylation of cytosine (5caC) in the CG dinucleotide in the E-box motif (CGCAG vertical bar GTG) increases binding of the Tcf3 vertical bar Ascl1 helix-loop-helix heterodimer 10-fold. *Biochemical and Biophysical Research Communications* 2014;449:248-55.

[244] Rozenberg JM, Shlyakhtenko A, Glass K, Rishi V, Myakishev MV, FitzGerald PC, et al. All and only CpG containing sequences are enriched in promoters abundantly bound by RNA polymerase II in multiple tissues. *BMC Genomics* 2008;9.

[245] Schubeler D. Function and information content of DNA methylation. *Nature* 2015;517:321-6.

[246] Kress C, Thomassin H, Grange T. Active cytosine demethylation triggered by a nuclear receptor involves DNA strand breaks. *Proc Natl Acad Sci U S A* 2006;103:11112-7.

[247] Ball MP, Li JB, Gao Y, Lee JH, LeProust EM, Park IH, et al. Targeted and genome-scale strategies reveal gene-body methylation signatures in human cells (vol 27, pg 361, 2009). *Nat Biotechnol* 2009;27:485-.

[248] Rauch TA, Wu XW, Zhong X, Riggs AD, Pfeifer GP. A human B cell methylome at 100-base pair resolution. *P Natl Acad Sci USA* 2009;106:671-8.

[249] Ziller M, Gu H, Müller F, Donaghey J, Tsai L, Kohlbacher O, et al. Charting a dynamic DNA methylation landscape of the human genome. *Nature* 2013;500:477-81.
[250] Kinoshita M, Numata S, Tajima A, Shimodera S, Ono S, Imamura A, et al. DNA Methylation Signatures of Peripheral Leukocytes in Schizophrenia. *Neuromol Med* 2013;15:95-101.
[251] Li S, Papale LA, Kintner DB, Sabat G, Barrett-Wilt GA, Cengiz P, et al. Hippocampal increase of 5-hmC in the glucocorticoid receptor gene following acute stress. *Behavioural Brain Research* 2015;286:236-40.
[252] Barrett LW, Fletcher S, Wilton SD. Regulation of eukaryotic gene expression by the untranslated gene regions and other non-coding elements. *Cellular and Molecular Life Sciences* 2012;69:3613-34.
[253] Kim M, Chen XW, Chin LJ, Paranjape T, Speed WC, Kidd KK, et al. Extensive sequence variation in the 3 ' untranslated region of the KRAS gene in lung and ovarian cancer cases. *Cell Cycle* 2014;13:1030-40.
[254] Martelli-Palomino G, Pancotto JA, Muniz YC, Mendes CT, Castelli EC, Massaro JD, et al. Polymorphic Sites at the 3 ' Untranslated Region of the HLA-G Gene Are Associated with Differential hla-g Soluble Levels in the Brazilian and French Population. *PLoS ONE* 2013;8.
[255] Zhang W, Winder T, Ning Y, Pohl A, Yang D, Kahn M, et al. A let-7 microRNA-binding site polymorphism in 3 '-untranslated region of KRAS gene predicts response in wild-type KRAS patients with metastatic colorectal cancer treated with cetuximab monotherapy. *Ann Oncol* 2011;22:104-9.
[256] Yang L, Li YY, Cheng M, Huang DS, Zheng J, Liu B, et al. A functional polymorphism at microRNA-629-binding site in the 3 '-untranslated region of NBS1 gene confers an increased risk of lung cancer in Southern and Eastern Chinese population. *Carcinogenesis* 2012;33:338-47.
[257] Jones PA. Functions of DNA methylation: islands, start sites, gene bodies and beyond. *Nature Reviews Genetics* 2012;13:484-92.
[258] Maunakea AK, Nagarajan RP, Bilenky M, Ballinger TJ, D'Souza C, Fouse SD, et al. Conserved role of intragenic DNA methylation in regulating alternative promoters. *Nature* 2010;466:253-U131.
[259] Cokus SJ, Feng SH, Zhang XY, Chen ZG, Merriman B, Haudenschild CD, et al. Shotgun bisulphite sequencing of the Arabidopsis genome reveals DNA methylation patterning. *Nature* 2008;452:215-9.
[260] Xie W, Barr CL, Kim A, Yue F, Lee AY, Eubanks J, et al. Base-Resolution Analyses of Sequence and Parent-of-Origin Dependent DNA Methylation in the Mouse Genome. *Cell* 2012;148:816-31.

[261] Vinson C, Chatterjee R. CG methylation. *Epigenomics-Uk* 2012;4:655-63.
[262] Lister R, Mukamel EA, Nery JR, Urich M, Puddifoot CA, Johnson ND, et al. Global Epigenomic Reconfiguration During Mammalian Brain Development. *Science* 2013;341:629.
[263] Zhang T, Ma J, Nie K, Yan J, Liu Y, Bacchi CE, et al. Hypermethylation of the tumor suppressor gene PRDM1/Blimp-1 supports a pathogenetic role in EBV-positive Burkitt lymphoma. *Blood Cancer J* 2014;4.
[264] Li SB, Zhu YF, Ma CG, Qiu ZH, Zhang XJ, Kang ZH, et al. Downregulation of EphA5 by promoter methylation in human prostate cancer. *Bmc Cancer* 2015;15.
[265] Zhang XY, Sun Q, Shan M, Niu M, Liu T, Xia BS, et al. Promoter Hypermethylation of ARID1A Gene Is Responsible for Its Low mRNA Expression in Many Invasive Breast Cancers. *Plos One* 2013;8.
[266] Wang P, Chen L, Zhang J, Chen H, Fan J, Wang K, et al. Methylation-mediated silencing of the miR-124 genes facilitates pancreatic cancer progression and metastasis by targeting Rac1. *Oncogene* 2014;33:514-24.
[267] Song I, Ha G, Kim J, Jeong S, Lee H, Kim Y, et al. Human ZNF312b oncogene is regulated by Sp1 binding to its promoter region through DNA demethylation and histone acetylation in gastric cancer. *Int J Cancer* 2011;129:2124-33.
[268] Søes S, Daugaard IL, Sørensen BS, Carus A, Mattheisen M, Alsner J, et al. Hypomethylation and increased expression of the putative oncogene ELMO3 are associated with lung cancer development and metastases formation. *Oncoscience* 2014;1.
[269] Liu H, Zhou Y, Boggs SE, Belinsky SA, Liu J. Cigarette smoke induces demethylation of prometastatic oncogene synuclein-gamma in lung cancer cells by downregulation of DNMT3B. *Oncogene* 2007;26:5900-10.
[270] Kamiyama H, Suzuki K, Maeda T, Koizumi K, Miyaki Y, Okada S, et al. DNA demethylation in normal colon tissue predicts predisposition to multiple cancers. *Oncogene* 2012:1-9.
[271] Borhani N, Manoochehri M, Gargari SS, Novin MG, Mansouri A, Omrani M. Decreased Expression of Proapoptotic Genes Caspase-8- and BCL2-Associated Agonist of Cell Death (BAD) in Ovarian Cancer. *Clinical Ovarian and Other Gynecologic Cancer* 2015.
[272] Bock C, Walter J, Paulsen M, Lengauer T. Inter-individual variation of DNA methylation and its implications for large-scale epigenome mapping. *Nucleic Acids Research* 2008;36.

[273] Nagaraj N, Mann M. Quantitative Analysis of the Intra- and Inter-Individual Variability of the Normal Urinary Proteome. *J Proteome Res* 2011;10:637-45.

[274] Hughes DA, Kircher M, He ZS, Guo S, Fairbrother GL, Moreno CS, et al. Evaluating intra- and inter-individual variation in the human placental transcriptome. *Genome Biology* 2015;16.

[275] Kamstra JH, Løken M, Aleström P, Legler J. Dynamics of DNA Hydroxymethylation in Zebrafish (Epub ahead of print). *Zebrafish* 2015.

[276] Almeida RD, Sottile V, Loose M, De Sousa PA, Johnson AD, Ruzov A. Semi-quantitative immunohistochemical detection of 5-hydroxymethylcytosine reveals conservation of its tissue distribution between amphibians and mammals. *Epigenetics* 2012;7:137-40.

[277] Neri F, Incamato D, Krepelova A, Rapelli S, Anselmi F, Parlato C, et al. Single-Base Resolution Analysis of 5-Formyl and 5-Carboxyl Cytosine Reveals Promoter DNA Methylation Dynamics. *Cell Reports* 2015;10:674-83.

[278] Alioui A, Wheldon L, Abakir A, Ferjentsik Z, Johnson A, Ruzov A. 5-Carboxylcytosine is localized to euchromatic regions in the nuclei of follicular cells in axolotl ovary. *Nucleus* 2012;3:565-9.

[279] Hur K, Cejas P, Feliu J, Moreno-Rubio J, Burgos E, Boland CR, et al. Hypomethylation of long interspersed nuclear element-1 (LINE-1) leads to activation of protooncogenes in human colorectal cancer metastasis. *Gut* 2014;63:635-46.

[280] Jin SG, Jiang Y, Quiu R, Rauch TA, Wang Y, Schackert G, et al. 5-Hydroxymethylcytosine Is Strongly Depleted in Human Cancers but Its Levels Do Not Correlate with IDH1 Mutations. *Cancer Research* 2011;71.

[281] Shpyleva S, Ivanovsky S, de Conti A, Melnyk S, Tryndyak V, Beland FA, et al. Cerebellar Oxidative DNA Damage and Altered DNA Methylation in the BTBR T plus tf/J Mouse Model of Autism and Similarities with Human Post Mortem Cerebellum. *PLoS ONE* 2014;9.

[282] Hu LL, Li Z, Cheng JD, Rao QH, Gong W, Liu MJ, et al. Crystal Structure of TET2-DNA Complex: Insight into TET-Mediated 5mC Oxidation. *Cell* 2013;155:1545-55.

[283] Gu TP, Guo F, Yang H, Wu HP, Xu GF, Liu W, et al. The role of Tet3 DNA dioxygenase in epigenetic reprogramming by oocytes. *Nature* 2011;477:606-U136.

[284] Wossidlo M, Nakamura T, Lepikhov K, Marques C, Zakhartchenko V, Boiani M, et al. 5-Hydroxymethylcytosine in the mammalian zygote is linked with epigenetic reprogramming. *Nat Commun* 2011;2.
[285] Capuano F, Mulleder M, Kok R, Blom HJ, Ralser M. Cytosine DNA Methylation Is Found in Drosophila melanogaster but Absent in Saccharomyces cerevisiae, Schizosaccharomyces pombe, and Other Yeast Species. *Analytical Chemistry* 2014;86:3697-702.
[286] Xia YY, Ding YB, Liu XQ, Chen XM, Cheng SQ, Li LB, et al. Racial/ethnic disparities in human DNA methylation. *Bba-Rev Cancer* 2014;1846:258-62.
[287] Subramanyam MA, Diez-Roux AV, Pilsner JR, Villamor E, Donohue KM, Liu YM, et al. Social Factors and Leukocyte DNA Methylation of Repetitive Sequences: The Multi-Ethnic Study of Atherosclerosis. *Plos One* 2013;8.
[288] Zhang F, Cardarelli R, Carroll J, Fulda K, Kaur M, Gonzales K, et al. Significant differences in global genomic DNA methylation by gender and race/ethnicity in peripheral blood. *Epigenetics-Us* 2011;6:623-9.
[289] Jenkins TG, Aston KI, Pflueger C, Cairns BR, Carrell DT. Age-Associated Sperm DNA Methylation Alterations: Possible Implications in Offspring Disease Susceptibility. *PLoS Genetics* 2014;10.
[290] Christensen B, Houseman A, Marsit C, Zheng S, Wrensch M, Wiemels J, et al. Aging and Environmental Exposures Alter Tissue-Specific DNA Methylation Dependent upon CpG Island Context. *Plos Genet* 2009;5:e1000602.
[291] Steegers-Theunissen RP, Obermann-Borst SA, Kremer D, Lindemans J, Siebel C, Steegers EA, et al. Periconceptional Maternal Folic Acid Use of 400 mu g per Day Is Related to Increased Methylation of the IGF2 Gene in the Very Young Child. *PLoS ONE* 2009;4.
[292] Ronn T, Volkov P, Davegardh C, Dayeh T, Hall E, Olsson AH, et al. A Six Months Exercise Intervention Influences the Genome-wide DNA Methylation Pattern in Human Adipose Tissue. *PLoS Genetics* 2013;9.
[293] Denham J, O'Brien B, Harvey J, Charchar F. Genome-wide sperm DNA methylation changes after 3 months of exercise training in humans. *Epigenomics-Uk* 2015.
[294] Essex M, Boyce W, Hertzman C, Lam L, Armstrong J, Neumann S, et al. Epigenetic Vestiges of Early Developmental Adversity: Childhood Stress Exposure and DNA Methylation in Adolescence. *Child Dev* 2013;84:58-75.

[295] Khulan B, Manning J, Dunbar D, Seckl J, Raikkonen K, Eriksson J, et al. Epigenomic profiling of men exposed to early-life stress reveals DNA methylation differences in association with current mental state. *Transl Psychiatry* 2014;4.

[296] Borghol N, Suderman M, McArdle W, Racine A, Hallett M, Pembrey M, et al. Associations with early-life socio-economic position in adult DNA methylation. *Int J Epidemiol* 2012;41:62-74.

[297] Ratel D, Ravanat JL, Berger F, Wion D. N6-methyladenine: the other methylated base of DNA. *Bioessays* 2006;28:309-15.

[298] Heyn H, Esteller M. An Adenine Code for DNA: A second life for N6-Methyladenine. *Cell* 2015;161:710-3.

[299] Zhang GQ, Huang H, Liu DR, Cheng Y, Liu X, Zhang W, et al. N6-Methyladenine DNA Modification in Drosophila. *Cell* 2015;161:893–906.

[300] Machado ACD, Zhou T, Rao S, Goel P, Rastogi C, Lazarovici A, et al. Evolving insights on how cytosine methylation affects protein-DNA binding. *Briefings in Functional Genomics* 2014.

[301] Schübeler D. Function and information content of DNA methylation. *Nature* 2015;517:321-6.

[302] Hendrich B, Bird A. Identification and characterization of a family of mammalian methyl-CpG binding proteins. *Molecular and Cellular Biology* 1998;18:6538-47.

[303] La HG, Ding B, Mishra GP, Zhou B, Yang HM, Bellizzi MD, et al. A 5-methylcytosine DNA glycosylase/lyase demethylates the retrotransposon Tos17 and promotes its transposition in rice. *P Natl Acad Sci USA* 2011;108:15498-503.

BIOGRAPHICAL SKETCH

Selcen Celik-Uzuner

Affiliation: Karadeniz Technical University, Department of Molcular Biology and Genetics,

Education: PhD, University of Sydney, Faculty of Medicine, Sydney, Australia MS, Ankara University, Biotechnology Institute, Ankara, Turkey; EdM, Gazi University, Faculty of Education, Biology Education, Ankara, Turkey; BS, Gazi University, Faculty of Science, Biology, Ankara, Turkey

Research and Professional Experience: My works mainly focus on DNA methylation in mammalian genome. My interests are "the faithful detection of cytosine modifications using immunofluorescence," "interactions between genes' methylation and environmental changes" and "regulation of DNA methylation in such cellular stress."

Publications Last Three Years:
Celik-Uzuner, S., and C. O'Neill (2016). "The sensitivity of 5-formylcytosine to doxorubicin regardless DNA damage," *Turkish Journal of Biology*, accepted, in publication process.
Celik, S., D. Akcora, T. Ozkan, N. Varol, S. Aydos and A. Sunguroglu (2015). "Methylation analysis of the DAPK1 gene in imatinib-resistant chronic myeloid leukemia patients." *Oncology Letters* 9(1): 399-404.
Celik, S. (2015). "Understanding the complexity of antigen retrieval of DNA methylation for immunofluorescence-based measurement and an approach to challenge." *Journal of Immunological Methods* 416: 1-16.

Celik, S., Y. Li and C. O'Neill (2015). "The effect of DNA damage on the pattern of immune-detectable DNA methylation in mouse embryonic fibroblasts." *Experimental Cell Research* 339(1): 20-34.

Celik, S., Y. Li and C. O'Neill (2014). "The Exit of Mouse Embryonic Fibroblasts from the Cell-Cycle Changes the Nature of Solvent Exposure of the 5 '-Methylcytosine Epitope within Chromatin." *Plos One* 9(4).

In: DNA Methylation
Editor: Kathleen Holland

ISBN: 978-1-53610-244-4
© 2016 Nova Science Publishers, Inc.

Chapter 3

TARGETING ABERRANT IGF2 EPIGENETICS AS A NOVEL ANTI-TUMOR APPROACH

Ji-Fan Hu[1,2], and Andrew R. Hoffman[2]*

[1]Stem Cell and Cancer Center, First Affiliated Hospital,
Jilin University, Changchun, Jilin Province, P. R. China
[2]VA Palo Alto Health Care System and
Stanford University Medical School, Palo Alto, US

ABSTRACT

Insulin-like growth factor 2 (*IGF2*) encodes a potent fetal mitogen that regulates cell proliferation, growth, differentiation and survival. In normal tissues, the gene is maternally imprinted, and its expression is epigenetically regulated by the coordination of differential allelic DNA methylation in the imprinting control region, CTCF/SUZ12-mediated intrachromosomal looping, and allelic chromatin histone modifications in the gene's promoters. Loss of *IGF2* imprinting has been observed in a variety of growth disorders and malignant tumors. This epigenetic mutation may provide investigators with novel biomarkers for early cancer detection, prediction, and prognosis. Targeting this tumor-specific epigenetic abnormality may represent a potential approach for the development of novel therapeutic anti-cancer strategies.

---

* Correspondence to: Ji-Fan Hu, M.D., Ph.D., E-mail: jifan@stanford.edu.

INTRODUCTION

There is a small subset of gene loci in the human genome that are subject to control by genomic imprinting [1, 2]. Imprinted genes are expressed in an allele-specific manner that is dependent on the origin of the parental chromosome [3]. Imprinted genes often play an important role in the development of embryos, placental formation, parenting behavior and metabolism. Abnormalities in the epigenetic regulation of these loci may cause aberrant expression of imprinted genes. Genetic abnormalities in imprinted loci have been implicated in the pathogenesis of several human disorders [4], including Prader-Willi/Angelman syndromes (15q11-q13), Beckwith–Wiedemann syndrome (BWS) (11p15.5), pseudoparahypothyroidism type 1b (20q13.3), Silver–Russell syndrome (SRS) (11p15.5), and transient neonatal diabetes mellitus type 1 (6q24). Beckwith–Wiedemann syndrome, which is characterized by fetal and postnatal overgrowth along with a high incidence of tumors, is associated with abnormal imprinting of the IGF2/H19 and KCNQ1-KCNQ10T1 loci on human chromosome 11p15.

Insulin-like growth factor 2 (*IGF2*) has been widely studied in the context of tumorigenesis. *IGF2* is located upstream of the noncoding RNA H19 gene in a small imprinted gene cluster at 11p15.5 [1, 5]. In most normal tissues except for liver and CNS, IGF2 is expressed exclusively from the paternal allele, whereas H19 is derived exclusively from the maternal allele. This epigenetic control mechanism may be lost in tumors, leading to the overexpression of *IGF2*.

IGF-II, the protein encoded by of IGF2, is a potent mitogenic growth factor that is essential for normal fetal development and growth. IGF-II exerts its growth-promoting and anti-apoptotic effects through the type 1 IGF receptor (IGF1R), which also mediates the action of IGF-I. Activation of IGF1R initiates PI3K/Akt and MAPK signaling cascades [6-10], resulting in cell proliferation and, in cancer, resistance to chemotherapeutic agents, radiation, and targeted therapies using tamoxifen and herceptin [6, 11, 12]. Abnormal activation of these IGF1R-directed pathways constitutes a target for the development of tumor-specific gene therapy [13].

In this chapter, we review loss of *IGF2* imprinting in human tumors, focusing especially on the mechanisms underlying this abnormality in epigenetic control. We also summarize recent progress on targeting this tumor-specific imprinting abnormality as a novel anti-cancer approach.

LOSS OF *IGF2* IMPRINTING AS A MOLECULAR HALLMARK IN TUMORS

The regulatory machinery underlying genomic imprinting establishes epigenetic modifications that lead to preferential allelic expression of a gene in a parent-of-origin-dependent manner. Biochemical imprints are reset during gametogenesis in every generation, and then re-established in the early stages of embryogenesis [14]. *IGF2* and *H19* are two reciprocally imprinted neighboring genes. Mice that carry the maternally-inherited null Igf2 gene are normal. In contrast, the paternally-inherited knockout allele causes severe growth retardation. On the other hand, *H19* is a maternally-expressed gene that encodes a long non-coding RNA that functions as a putative tumor suppressor [15] whose precise biological role remains unresolved.

IGF2 is imprinted in a tissue-specific, promoter-specific, and development-specific manner. The expression of *IGF2* is controlled by four promoters. Promoter 1 (hP1), located immediately downstream of the insulin gene, is not imprinted and is normally biallelically expressed [16]. In most cells, hP1 is expressed at very low or undetectable levels, and it does not contribute substantially to *IGF2* expression. In contrast, promoters 2-4 (hP2-hP4), clustered distally from hP1, are imprinted, with expression only from the paternal allele [16]. Mouse *Igf2* contains three promoters (mP1-mP3) that are structurally homologous to the human counterpart hP2-hP4. They are also imprinted in all tissues examined, except for CNS, where both the parental alleles are expressed [17] in both human and mouse, possibly reflecting a compensatory mechanism for preventing low growth factor supply in the CNS.

In tumor tissues, however, this regulatory system is dysfunctional and both parental copies of the *IGF2* gene may become fully expressed [18-20], leading to abnormally high IGF-II production. Reactivation of the normally-suppressed (imprinted) maternal allele, known as loss of imprinting (LOI), is a hallmark of many human tumors, especially childhood tumors [1, 2] and cancer stem cells[21]. Existing evidence shows that the imprinting of the three *IGF2* clustered promoters (hP2-hP4) is coordinately regulated. Using promoter-allelic PCR, we have observed that in tumors where there is LOI for *IGF2*, expression from all three promoters hP2-hP4 is biallelic [22].

Aberrant *IGF2* imprinting occurs in a variety of human malignancies, including hepatoma, lung cancer, breast cancer, colorectal cancer, leiomyosarcoma, osteosarcoma, leukemia, and Wilms' tumor [18-20, 23-29]. Over-expression of *IGF2* mRNA may lead to increased IGF-II peptide which can promote tumor growth via autocrine and/or paracrine interactions that enhance cell growth and cancer stem cell self-renewal. *IGF2*-overexpressing tumors frequently display loss of *PTEN*, and they are frequently highly proliferative, exhibiting strong staining for phospho-Akt. These LOI tumors belong to a subclass of neoplasms characterized by poor survival [30]. Detection of *IGF2* LOI in circulating white blood cells represents a valuable biomolecular marker for predicting individuals with high risk for colorectal cancer [31].

Biallelic *IGF2* expression is as an early event in tumorigenesis both in animal models [32-35] and in human studies [36-40]. Development of β-cell carcinomas in transgenic mice carrying the insulin regulatory region/ST-40 large T-antigen requires the abnormal activation of the maternal *Igf2* allele as the second signal in the tumor pathway [32]. Biallelic activation of *Igf2* coincides with the switch to the hyperproliferative stage in preneoplastic foci and tumors [35], while homologous deletion of the *Igf2* gene significantly reduces tumor burden [32, 33].

By introducing DNA hypermethylation in the gene promoter using methylated hairpin oligonucleotides, we demonstrated transcriptional suppression of *IGF2* both *in vitro* and *in vivo* [41]. Silencing of *IGF2* reduced the growth of implanted human hepatocarcinomas and prolonged lifespan in an animal model [42, 43]. Feinberg and colleagues have shown that loss of *IGF2* imprinting in peripheral blood leukocytes may provide a potential biomarker in diagnosing individuals with high risk of colorectal cancer [38, 44], and that tumor phenotype in Apc+/Min mice can be modified simply by altering *IGF2* epigenotype [45], reinforcing the concept that *IGF2* plays a role in cancer.

EPIGENETIC MECHANISMS UNDERLYING LOSS OF *IGF2* IMPRINTING IN TUMORS

For a gene to be imprinted, the cellular transcriptional machinery must be able to distinguish between two parental alleles. The biochemical imprint must be epigenetic, stably inherited throughout embryo development, and be reversible such that both alleles can be appropriately reset during gametogenesis [46-51]. DNA methylation plays an important role in the establishment and maintenance of genomic imprinting. Gene knockout mice that lack DNA methyltransferase Dnmt1 show aberrant allelic expression of imprinted genes [52].

It is interesting to note that all three *IGF2* imprinted promoters (hP2-hP4), although being rich in CpG islands, are not differentially methylated in the two parental alleles [53-55], thus excluding the control of allelic expression by DNA methylation in the gene promoter. However, both human and murine *IGF2* genes contain several differentially methylated DNA regions (DMRs) [50, 56-60], including DMR$_0$, DMR$_1$, DMR$_2$, and CTCF DMR. DMRs are characterized by DNA methylation on one of the two parental alleles. Such *cis* elements contain differentially methylated CpG-rich repeats on parental alleles, which are postulated to function as the so-called "imprinting signal" that guides allelic gene silencing at imprinted loci. Aberrant DNA methylation has been reported in human tumors, especially at CTCF binding sites [29, 31, 61-63].

It was initially thought that *IGF2* and *H19* expression would be tightly coordinated, reciprocally controlled by an "enhancer competition" mechanism [64]. Deletion of the *H19* enhancers alone [65] or the *H19* gene plus its enhancers [66] changes allelic expression of both genes, indicating that *H19* and *IGF2* utilize the same enhancers that act differently on each parental chromosome [67]. The finding of a CTCF "boundary insulator" at the CTCF DMR in an imprinting control region (ICR) between *H19* and *IGF2* [68, 69] further delineated the role of both *cis* modifications and *trans* factors in controlling allelic expression. Differential methylation of the CTCF DMR in the ICR is established during spermatogenesis. This ICR contains seven CCCTC-binding factor (CTCF) binding sites in human and four in mouse. The ICR, which regulates the reciprocal expression of *IGF2* and *H19*, is located 2 kb upstream of *H19* and is methylated on the paternal but not the maternal allele. CTCF binding creates a physical boundary that blocks the interaction of downstream enhancers with the remote IGF2 promoters and silences the maternal allele [50, 56, 57, 64, 68-73]. Hypermethylation of the paternal ICR, however, abrogates the binding of CTCF and thus allows the exclusive expression of *H19* from the maternal allele and *IGF2* from the paternal allele, ensuring the reciprocal imprinting of these two neighboring genes.

However, the molecular events involved in the loss of *IGF2* imprinting (LOI) in human tumors have not been fully characterized. In human tumors, loss of *IGF2* imprinting occurs long after the imprint had been set as opposed to some genetic diseases and animal models in which imprinting never developed. In a subset of Wilms' tumors, colon cancers, and osteosarcomas, *IGF2* LOI is accompanied by aberrant methylation of the maternal *H19 DMR* and silencing of the maternal *H19* allele [29, 61, 62, 74]. In some tumors, *IGF2* LOI is related to aberrant CpG methylation within a region corresponding to the mouse *IGF2 DMR$_0$* [31].

A series of studies from our lab suggest that CTCF may not only function as a physical insulator as originally suggested, but also actively participate in the regulation of the imprinted *IGF2* allele [53, 75-77]. Chromosome configuration capture (3C) has been used to reveal the role of CTCF in the formation of long-range chromosomal loops [73, 78, 79]. CTCF acts as a tethering protein, serving as a molecular glue to secure long range intrachromosomal [53, 78] and interchromosomal [80] interactions. CTCF-mediated chromatin looping brings the ICR and the *IGF2* promoters into close contact, where the polycomb repressive complex 2 (PCR2) is recruited via SUZ12, inducing allelic H3K27 methylation and gene silencing [53, 75]. Thus, an intrachromosomal scaffold built with CTCF guides the imprinting signal to the remote ICR to establish the suppressive histone code in the distant *IGF2* promoters.

Loss of *IGF2* imprinting is associated with loss of the normal intrachromosomal loop in human tumors and this could be the result of an abnormality in any of the steps during the formation of the CTCF-PRC2-ICR-promoter intrachromosomal complex. One defect could be altered CpG DNA methylation in the ICR, where the imprinting signal resides in mouse [68, 69]. For example, hypermethylation at both parental ICRs could prevent the binding of CTCF and consequently the failure to form the intrachromosomal loop complex. A detailed analysis of DNA methylation at the *IGF2* locus in human cancers, however, shows that *IGF2* LOI was not necessarily linked to, and may be independent of, epigenetic marks in the various DMRs, including the ICR[55]. In some tumors, *IGF2* LOI persists even when the ICR maintains its normally differentially methylated state. In some tumors, *IGF2* imprinting is not lost yet there are abnormal epigenetic modifications, such as hypomethylation or hypermethylation, at CTCF binding sites.

We have shown that aberrant *IGF2* imprinting in tumor cell lines may be related to the loss of the histone H3K27 methylation suppressive mark in the gene promoter [54]. SUZ12 is downregulated in LOI tumor cell lines. In the absence of SUZ12 binding, CTCF is unable to orchestrate a long range intrachromosomal loop that juxtaposes the ICR close to the gene promoters, enabling the establishment of H3K27 methylation suppression through EZH2, a methyltransferase component of PRC2. In the absence of H3K27 methylation, the maternal allele becomes activated, leading to biallelic expression of *IGF2*.

Aberrant biallelic expression of *IGF2* in human tumors is associated with the loss of the CTCF-orchestrated intrachromosomal complex, which is required for the recruitment of the PRC2 via the interaction with SUZ12 [53, 75]. Without formation of the intrachromosomal complex, H3K27 methyltransferase

EZH2 cannot be guided to the maternal *IGF2* promoters, where it establishes the suppressive epigenotype. The H3K27 methylation-free promoters then become activated in a similar fashion as in the paternal promoters. Correct looping is ensured by specific methylation patterns of the DMRs, and in particular by CTCF, which binds only unmethylated DNA in the central ICR.

CORRECTION OF ABERRANT IMPRINTING BY POLYCOMB REPRESSIVE COMPLEX 2 DOCKING FACTOR SUZ12

Resetting of epigenetic modifications by epigenetic reprogramming plays a critical role in the dedifferentiation of the terminally differentiated nucleus. We demonstrate that loss of *IGF2* imprinting in tumor cells can be corrected or normalized by "epigenetic reprogramming" when nuclei from tumor cells are transferred into a cellular environment where *IGF2* imprinting is normally maintained [55]. Prior to the transfer of nuclei from LOI cells into the cytoplasm of maintenance of imprinting (MOI) cytoplasm, the four human tumor cell lines (WTCL, SKNEP, HRT18, and H522) expressed *IGF2* biallelically. In reconstructed cybrids, however, monoallelic expression of *IGF2* was observed, indicating correction of abnormal *IGF2* imprinting by nuclear reprogramming [55]. Similarly, treatment of cells with cycloheximide that depletes the cells of the putative imprinting factors recapitulates the loss of imprinting in tumors [55]. Clearly, the cytoplasm of normal fibroblasts contains sufficient imprinting maintenance machinery needed to reset the *IGF2* imprint in nuclei derived from tumor cells. The intact *IGF2* imprinting system in normal cytoplasm corrects aberrant imprinting by providing *trans* imprinting factors that were missing in tumor cells.

Several *trans* factors have been implicated in the regulation of imprinting. PRC2, containing core proteins *SUZ12* (suppressor of zeste 12), *EED* (embryonic ectoderm development) and *EZH2* (enhancer of zeste homologue 2), catalyzes the di- and tri-methylation of histone H3 at lysine 27 [81-83]. PcG proteins have been shown to play key roles in stem cell biology [84, 85], cancer epigenetics [86, 87], X inactivation [88], imprinting [89] and multicellular development in plant and animal systems [90, 91]. Our recent studies indicate a role for polycomb repressive complex 2 (PRC2) [53, 75-77], particularly its docking factor *SUZ12* [54], in the regulation of *IGF2* allelic expression in many tissues. To examine the role of chromatin factors in the maintenance of *IGF2* imprinting, we used Western blotting to compare the abundance of CTCF and

SUZ12 proteins between the maintenance of imprinting and LOI cells. We did not detect any differences in CTCF abundance between the MOI and LOI cell lines. PRC2 docking factor SUZ12, however, was dramatically downregulated in all LOI tumor cell lines [54]. Similarly, the abundance of SUZ12 protein was also low in several *IGF2* LOI cells induced by cycloheximide treatment. Thus, downregulation of *SUZ12* expression may play an important role in the loss of *IGF2* imprinting in these tumors.

Exactly how PRC2 is recruited to chromatin is not clear [54, 92]. By virally expressing the cDNA of *SUZ12* in two LOI human colon cancer cell lines, we demonstrate that *IGF2* expression is restored to normal monoallelic expression. Virally expressed SUZ12 protein binds to the *IGF2* promoters and coordinates with CTCF to orchestrate a long range intrachromosomal loop, leading to histone H3K27 methylation in the *IGF2* promoters and restoring monoallelic expression of *IGF2* in tumor cells [93]. The critical involvement of SUZ12 in maintaining the normal monoallelic expression of *IGF2* suggests that the PRC2 regulatory pathway may be dysfunctional in human tumors in which *IGF2* is biallelically expressed.

AN ONCOLYTIC VIRAL THERAPY USING THE *IGF2/H19* ENHANCER-INSULATOR-PROMOTER SYSTEM

If *IGF2* LOI is found only in the tumor, then we can use the LOI epigenotype as a molecular target for developing a novel form of anti-cancer therapy. To target *IGF2* LOI in tumors, we developed a "mini-imprinting cassette" strategy. We constructed a LOI-oncolytic adenovirus containing the *IGF2/H19* enhancer-ICR-promoter cassette (Figure 1) [94], which mimics the imprinting control system. In this mini-imprinting cassette, the *IGF2* promoter is controlled by an upstream H19 enhancer. An ICR fragment containing six CTCF binding sites is inserted between the enhancer and the promoter to insulate its activity as previously reported [68, 69].

A therapeutic gene, like Ad5 early region gene 1A (E1A) or diphtheria A toxin (DT-A), is placed downstream of the ICR-controlled *IGF2* promoter. With this design, expression of the therapeutic gene will be under the tight control of the host imprinting machinery.

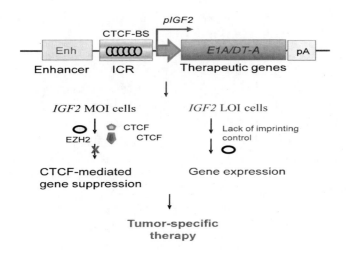

Figure 1. Imprinting-mediated tumor therapy. Enh: *H19* downstream enhancer; ICR: imprinting control region; CTCF-BS: CTCF factor binding sites; pA: TK poly A single site; p*IGF2*: *IGF2* promoter; E1A: adenoviral replication gene; DT-A: diphtheria toxin A gene; MOI: maintenance of imprinting; LOI: loss of imprinting.

In normal cells, the imprinting control system is functional. CTCF binds to the unmethylated ICR and insulates the activity of the nearby promoter [68, 69]. The binding of CTCF recruits polycomb repressive complex 2 (PRC2) via the docking factor SUZ12, which introduces a *de novo* histone H3-K27 methylation in the downstream promoter [54, 92]. As a result of this normal imprinting control, the downstream *IGF2* promoter is suppressed and the therapeutic toxic gene is silenced, thus sparing normal cells. In LOI tumors, however, the dysfunctional imprinting control mechanism, like downregulation of PRC2 docking factor SUZ12 and abnormal DNA methylation in the ICR [53, 54], will fail to insulate the promoter. This leads to the LOI-specific activation of E1A or DT-A, which specifically lyses or kills tumor cells.

The proof-of-concept of this epigenetic approach was tested using an EGFP reporter gene under imprinting control [94]. The EGFP reporter was inserted downstream of the mini-imprinting cassette. It is assumed that the EGFP reporter would be expressed in those LOI tumor cell lines but not in the MOI cells. The packaged adenovirus was used to infect tumor cells. We noted the strong fluorescence of the expressed EGFP in HRT-18 and HT-29 tumors that show *IGF2* LOI, while the negative or very weak fluorescence was observed in HCT-116, MCF-7 and GES-1 cells that maintain normal *IGF2* imprinting, regardless of the multiplicity of infection or the prolonged infection time.

The therapeutic effect of this imprinting control system was then tested using diphtheria toxin DT-A. DT-A inhibits protein synthesis in susceptible cells by binding directly to NAD+ and catalyzing the transfer of ADP ribose from NAD+ to elongation factor 2 and irreversibly inhibiting its activity [95]. Tumor promoter-controlled DT-A has been used as a cancer gene therapy to achieve efficacious killing of cancer cells [96, 97]. By putting DT-A under the control of the imprinting machinery, the therapeutic adenovirus will specifically kill or lyse tumor cells that show aberrant *IGF2* imprinting.

After transfection with the enhancer-ICR-promoter-DT-A adenovirus, both RT-PCR and Western blot assays detected the expression of DT-A in three LOI tumor cells (HT29, HCT8, and H522). No DT-A products were detected in GES1 and MCF7 cells that show normal *IGF2* imprinting. The expressed DT-A toxin caused specific cytotoxicity and apoptosis in three LOI tumor cells. Similarly, this epigenetic therapy reduced the tumor burden in animals with transplanted tumor grafts [94].

Alternatively, this epigenetic approach can be developed as an imprinting-controlled oncolytic therapy using adenoviral E1A as a therapeutic gene [98]. E1A encodes the first adenoviral proteins that are produced upon infection and function primarily as the activator of transcription of the other viral early gene products (17). Expression of E1A is critical for the replication of adenoviruses [99, 100]. The E1A proteins alter the host cell to allow more efficient production of viral progeny. In addition, E1A may induce apoptosis (18), and ElA may enhance sensitivity to chemotherapeutics and radiation. When controlled by a cancer-specific promoter, expression of E1A yields a tumor cell-selective, conditionally replicating adenovirus that specifically lyses tumor cells, while normal cells are spared [99, 100]. Imprinting-specific activation of E1A would allow the growth-defective adenovirus to replicate and finally lyse tumor cells.

We constructed the LOI-oncolytic vector that express E1A protein only in cells where there was a loss of *IGF2* imprinting. The therapeutic effect of the enhancer-ICR-promoter-E1A model was tested in two *IGF2* LOI colon cancer cells (HRT-18 and HT-29) and three *IGF2* MOI cells (colon cancer cell HCT-116, breast cancer cell MCF-7, and gastric epithelial cell GES-1). After infection, the imprinting-controlled E1A oncolytic adenovirus replicates selectively with high efficiency only in cells with LOI. The imprinting-dependent induction of oncolysis and apoptosis is observed in LOI-tumor cells, but not in MOI cell lines [98, 101]. Similarly, in HRT-18 and HT-29 xenograft mouse models, the imprinting-controlled oncolytic adenovirus increased the apoptotic index, leading to significant inhibition of tumor growth [98, 101] and prolonged the tumor survival [101]. These studies suggest that the imprinting-controlled E1A vector has therapeutic potential for colon cancer.

CONCLUSION

Abnormalities in the epigenetic control of *IGF2* imprinting is a hallmark of tumorigenesis. We still do not fully understand the mechanisms underlying aberrant imprinting that occurs in some tumors. In addition to the regulation by PcG proteins and CTCF, other chromatin modifying factors like long noncoding RNAs, small RNAs and miRNAs may also be involved in local chromatin organization that determines the allelic expression of *IGF2*. The roles of these factors in the pathogenesis of imprinting disorders is currently being explored.

Progress in these fields may provide new diagnostic tools for the clinic. For example, monitoring abnormal *IGF2* imprinting in peripheral blood cells may be a useful biomarker for predicting colon cancer risk [38, 44].

In addition, the *IGF2* LOI system may offer an interesting approach to contribute to future development of imprinting-based therapies. However, it should also be noted that aberrant *IGF2* imprinting occurs only in a subgroup of cancer patients. It will also be interesting to explore if this therapeutic approach can be extended to tumors with subtle epigenetic abnormalities before they affect imprinting.

ACKNOWLEDGMENTS

This work was supported by NIH grant (1R43 CA103553-01), Department of Defense Grant (W81XWH-04-1-0597), California Institute of Regenerative Medicine (CIRM) grant (RT2-01942), NSFC grant (81272294, 31430021) to J.F.H; and NIH grant (GM09031) and a Merit Review from the Medical Research Service of the Department of Veterans Affairs to A.R.H.

REFERENCES

[1] Bergman D, Halje M, Nordin M, Engstrom W. Insulin-like growth factor 2 in development and disease: a mini-review. *Gerontology*. 2013;59: 240-249.

[2] Ribarska T, Bastian KM, Koch A, Schulz WA. Specific changes in the expression of imprinted genes in prostate cancer--implications for cancer progression and epigenetic regulation. *Asian J Androl.* 2012;14: 436-450.

[3] Hoffman AR, Vu TH, Hu J. Mechanisms of genomic imprinting. *Growth Horm IGF Res.* 2000;10: S18-19.

[4] Skaar DA, Li Y, Bernal AJ, Hoyo C, Murphy SK, Jirtle RL. The human imprintome: regulatory mechanisms, methods of ascertainment, and roles in disease susceptibility. *ILAR J.* 2012;53: 341-358.

[5] Riccio A, Sparago A, Verde G, et al. Inherited and Sporadic Epimutations at the IGF2-H19 locus in Beckwith-Wiedemann syndrome and Wilms' tumor. *Endocr Dev.* 2009;14: 1-9.

[6] Chapuis N, Tamburini J, Cornillet-Lefebvre P, et al. Autocrine IGF-1/IGF-1R signaling is responsible for constitutive PI3K/Akt activation in acute myeloid leukemia: therapeutic value of neutralizing anti-IGF-1R antibody. *Haematologica.* 2010;95: 415-423.

[7] Pollak M. The insulin and insulin-like growth factor receptor family in neoplasia: an update. *Nat Rev Cancer.* 2012;12: 159-169.

[8] Pierre-Eugene C, Pagesy P, Nguyen TT, et al. Effect of insulin analogues on insulin/IGF1 hybrid receptors: increased activation by glargine but not by its metabolites M1 and M2. *PLoS One.* 2012;7: e41992.

[9] Huang GS, Brouwer-Visser J, Ramirez MJ, et al. Insulin-like growth factor 2 expression modulates Taxol resistance and is a candidate biomarker for reduced disease-free survival in ovarian cancer. *Clin Cancer Res.* 2010;16: 2999-3010.

[10] Danielsen SA, Eide PW, Nesbakken A, Guren T, Leithe E, Lothe RA. Portrait of the PI3K/AKT pathway in colorectal cancer. *Biochim Biophys Acta.* 2015;1855: 104-121.

[11] Grandage VL, Gale RE, Linch DC, Khwaja A. PI3-kinase/Akt is constitutively active in primary acute myeloid leukaemia cells and regulates survival and chemoresistance via NF-kappaB, Mapkinase and p53 pathways. *Leukemia.* 2005;19: 586-594.

[12] Xu Q, Simpson SE, Scialla TJ, Bagg A, Carroll M. Survival of acute myeloid leukemia cells requires PI3 kinase activation. *Blood.* 2003;102: 972-980.

[13] Rieder S, Michalski CW, Friess H, Kleeff J. Insulin-like growth factor signaling as a therapeutic target in pancreatic cancer. *Anticancer Agents Med Chem.* 2011;11: 427-433.

[14] Barlow DP. Genomic imprinting: a mammalian epigenetic discovery model. *Annu Rev Genet.* 2011;45: 379-403.

[15] Hao Y, Crenshaw T, Moulton T, Newcomb E, Tycko B. Tumour-suppressor activity of H19 RNA. *Nature.* 1993;365: 764-767.
[16] Vu TH, Hoffman AR. Promoter-specific imprinting of the human insulin-like growth factor-II gene. *Nature.* 1994;371: 714-717.
[17] Hu J, Vu T, Hoffman A. Differential biallelic activation of three insulin-like growth factor II promoters in the mouse central nervous system. *Mol Endocrinol.* 1995;9: 628-636.
[18] Ogawa O, Eccles MR, Szeto J, et al. Relaxation of insulin-like growth factor II gene imprinting implicated in Wilms' tumour. *Nature.* 1993;362: 749-751.
[19] Rainier S, Johnson LA, Dobry CJ, Ping AJ, Grundy PE, Feinberg AP. Relaxation of imprinted genes in human cancer. *Nature.* 1993;362: 747-749.
[20] Feinberg AP. Genomic imprinting and gene activation in cancer. *Nat Genet.* 1993;4: 110-113.
[21] Hofmann WK, Takeuchi S, Frantzen MA, Hoelzer D, Koeffler HP. Loss of genomic imprinting of insulin-like growth factor 2 is strongly associated with cellular proliferation in normal hematopoietic cells. *Exp Hematol.* 2002;30: 318-323.
[22] Vu TH, Nguyen AH, Hoffman AR. Loss of IGF2 imprinting is associated with abrogation of long-range intrachromosomal interactions in human cancer cells. *Hum Mol Genet.* 2009;19: 901-919.
[23] Zhang L, Zhou W, Velculescu VE, et al. Gene expression profiles in normal and cancer cells. *Science.* 1997;276: 1268-1272.
[24] Sohda T, Iwata K, Soejima H, Kamimura S, Shijo H, Yun K. In situ detection of insulin-like growth factor II (IGF2) and H19 gene expression in hepatocellular carcinoma. *J Hum Genet.* 1998;43: 49-53.
[25] Takeda S, Kondo M, Kumada T, et al. Allelic-expression imbalance of the insulin-like growth factor 2 gene in hepatocellular carcinoma and underlying disease. *Oncogene.* 1996;12: 1589-1592.
[26] Boulle N, Logie A, Gicquel C, Perin L, Le Bouc Y. Increased levels of insulin-like growth factor II (IGF-II) and IGF- binding protein-2 are associated with malignancy in sporadic adrenocortical tumors. *J Clin Endocrinol Metab.* 1998;83: 1713-1720.
[27] Wu MS, Wang HP, Lin CC, et al. Loss of imprinting and overexpression of IGF2 gene in gastric adenocarcinoma. *Cancer Lett.* 1997;120: 9-14.
[28] Kim KS, Lee YI. Biallelic expression of the H19 and IGF2 genes in hepatocellular carcinoma. *Cancer Lett.* 1997;119: 143-148.

[29] Ulaner GA, Vu TH, Li T, et al. Loss of imprinting of *Igf2* and *H19* in osteosarcoma is accompanied by reciprocal methylation changes of a CTCF-binding site. *Hum Mol Genet*. 2003;12: 535-549.

[30] Soroceanu L, Kharbanda S, Chen R, et al. Identification of IGF2 signaling through phosphoinositide-3-kinase regulatory subunit 3 as a growth-promoting axis in glioblastoma. *Proc Natl Acad Sci USA*. 2007;104: 3466-3471.

[31] Cui H, Onyango P, Brandenburg S, Wu Y, Hsieh CL, Feinberg AP. Loss of imprinting in colorectal cancer linked to hypomethylation of H19 and IGF2. *Cancer Res*. 2002;62: 6442-6446.

[32] Christofori G, Naik P, Hanahan D. A second signal supplied by insulin-like growth factor II in oncogene-induced tumorigenesis. *Nature*. 1994;369: 414-418.

[33] Haddad R, Held WA. Genomic imprinting and Igf2 influence liver tumorigenesis and loss of heterozygosity in SV40 T antigen transgenic mice. *Cancer Res*. 1997;57: 4615-4623.

[34] Hu JF, Cheng Z, Chisari FV, Vu TH, Hoffman AR, Campbell TC. Repression of hepatitis B virus (HBV) transgene and HBV-induced liver injury by low protein diet. *Oncogene*. 1997;15: 2795-2801.

[35] Christofori G, Naik P, Hanahan D. Deregulation of both imprinted and expressed alleles of the insulin- like growth factor 2 gene during beta-cell tumorigenesis. *Nat Genet*. 1995;10: 196-201.

[36] Cui H, Horon IL, Ohlsson R, Hamilton SR, Feinberg AP. Loss of imprinting in normal tissue of colorectal cancer patients with microsatellite instability. *Nat Med*. 1998;4: 1276-1280.

[37] Okamoto K, Morison IM, Taniguchi T, Reeve AE. Epigenetic changes at the insulin-like growth factor II/H19 locus in developing kidney is an early event in Wilms tumorigenesis. *Proc Natl Acad Sci USA*. 1997;94: 5367-5371.

[38] Cruz-Correa M, Cui H, Giardiello FM, et al. Loss of imprinting of insulin growth factor II gene: a potential heritable biomarker for colon neoplasia predisposition. *Gastroenterology*. 2004;126: 964-970.

[39] Yuan E, Li CM, Yamashiro DJ, et al. Genomic profiling maps loss of heterozygosity and defines the timing and stage dependence of epigenetic and genetic events in Wilms' tumors. *Mol Cancer Res*. 2005;3: 493-502.

[40] Wang WH, Duan JX, Vu TH, Hoffman AR. Increased expression of the insulin-like growth factor-II gene in Wilms' tumor is not dependent on loss of genomic imprinting or loss of heterozygosity. *J Biol Chem*. 1996;271: 27863-27870.

[41] Hoffman AR, Hu JF. Directing DNA Methylation to Inhibit Gene Expression. *Cell Mol Neurobiol.* 2006;26: 425-438.

[42] Yao XM, Hu JF, Daniels M, et al. A methylated oligonucleotide inhibits IGF2 expression and enhances survival in a model of hepatocellular carcinoma. *J Clin Invest.* 2003;111: 265-273.

[43] Yao XM, Hu JF, Daniels M, et al. A novel orthotopic tumor model to study growth factors and oncogenes in hepatocarcinogenesis. *Clin Cancer Res.* 2003;9: 2719-2726.

[44] Cui H, Cruz-Correa M, Giardiello FM, et al. Loss of IGF2 imprinting: a potential marker of colorectal cancer risk. *Science.* 2003;299: 1753-1755.

[45] Sakatani T, Kaneda A, Iacobuzio-Donahue CA, et al. Loss of imprinting of Igf2 alters intestinal maturation and tumorigenesis in mice. *Science.* 2005;307: 1976-1978.

[46] Tycko B. DNA methylation in genomic imprinting. *Mutat Res.* 1997;386: 131-140.

[47] Bartolomei MS, Tilghman SM. Genomic imprinting in mammals. *Annu Rev Genet.* 1997;31: 493-525.

[48] Constancia M, Pickard B, Kelsey G, Reik W. Imprinting mechanisms. *Genome research.* 1998;8: 881-900.

[49] Reik W, Dean W. DNA methylation and mammalian epigenetics. *Electrophoresis.* 2001;22: 2838-2843.

[50] Mann JR, Szabo PE, Reed MR, Singer-Sam J. Methylated DNA sequences in genomic imprinting. *Crit Rev Eukaryot Gene Expr.* 2000;10: 241-257.

[51] Brannan CI, Bartolomei MS. Mechanisms of genomic imprinting. *Curr Opin Genet Dev.* 1999;9: 164-170.

[52] Li E, Beard C, Jaenisch R. Role for DNA methylation in genomic imprinting. *Nature.* 1993;366: 362-365.

[53] Li T, Hu JF, Qiu X, et al. CTCF regulates allelic expression of Igf2 by orchestrating a promoter-polycomb repressive complex-2 intrachromosomal loop. *Mol Cell Biol.* 2008;28: 6473-6482.

[54] Li T, Chen H, Li W, et al. Promoter histone H3K27 methylation in the control of IGF2 imprinting in human tumor cell lines. *Hum Mol Genet.* 2014;23: 117-128.

[55] Chen HL, Li T, Qiu XW, et al. Correction of aberrant imprinting of IGF2 in human tumors by nuclear transfer-induced epigenetic reprogramming. *EMBO J.* 2006;25: 5329-5338.

[56] Arney KL. H19 and Igf2 - enhancing the confusion? *Trends Genet.* 2003;19: 17-23.

[57] Sasaki H, Ishihara K, Kato R. Mechanisms of Igf2/H19 imprinting: DNA methylation, chromatin and long-distance gene regulation. *J Biochem.* 2000;127: 711-715.

[58] Thorvaldsen JL, Duran KL, Bartolomei MS. Deletion of the H19 differentially methylated domain results in loss of imprinted expression of H19 and Igf2. *Genes Dev.* 1998;12: 3693-3702.

[59] Szabo PE, Tang SH, Silva FJ, Tsark WM, Mann JR. Role of CTCF binding sites in the Igf2/H19 imprinting control region. *Mol Cell Biol.* 2004;24: 4791-4800.

[60] Srivastava M, Hsieh S, Grinberg A, Williams-Simons L, Huang SP, Pfeifer K. H19 and Igf2 monoallelic expression is regulated in two distinct ways by a shared cis acting regulatory region upstream of H19. *Genes Dev.* 2000;14: 1186-1195.

[61] Cui H, Niemitz EL, Ravenel JD, et al. Loss of imprinting of insulin-like growth factor-II in Wilms' tumor commonly involves altered methylation but not mutations of CTCF or its binding site. *Cancer Res.* 2001;61: 4947-4950.

[62] Nakagawa H, Chadwick RB, Peltomaki P, Plass C, Nakamura Y, de La Chapelle A. Loss of imprinting of the insulin-like growth factor II gene occurs by biallelic methylation in a core region of H19-associated CTCF-binding sites in colorectal cancer. *Proc Natl Acad Sci USA.* 2001;98: 591-596.

[63] Lopes S, Lewis A, Hajkova P, et al. Epigenetic modifications in an imprinting cluster are controlled by a hierarchy of DMRs suggesting long-range chromatin interactions. *Hum Mol Genet.* 2003;12: 295-305.

[64] Bartolomei MS, Webber AL, Brunkow ME, Tilghman SM. Epigenetic mechanisms underlying the imprinting of the mouse H19 gene. *Genes Dev.* 1993;7: 1663-1673.

[65] Leighton PA, Saam JR, Ingram RS, Stewart CL, Tilghman SM. An enhancer deletion affects both H19 and Igf2 expression. *Genes Dev.* 1995;9: 2079-2089.

[66] Leighton PA, Ingram RS, Eggenschwiler J, Efstratiadis A, Tilghman SM. Disruption of imprinting caused by deletion of the H19 gene region in mice. *Nature.* 1995;375: 34-39.

[67] Abramowitz LK, Bartolomei MS. Genomic imprinting: recognition and marking of imprinted loci. *Curr Opin Genet Dev.* 2012;22: 72-78.

[68] Bell AC, Felsenfeld G. Methylation of a CTCF-dependent boundary controls imprinted expression of the Igf2 gene. *Nature.* 2000;405: 482-485.

[69] Hark AT, Schoenherr CJ, Katz DJ, Ingram RS, Levorse JM, Tilghman SM. CTCF mediates methylation-sensitive enhancer-blocking activity at the H19/Igf2 locus. *Nature.* 2000;405: 486-489.
[70] Kanduri C, Pant V, Loukinov D, et al. Functional association of CTCF with the insulator upstream of the H19 gene is parent of origin-specific and methylation-sensitive. *Curr Biol.* 2000;10: 853-856.
[71] Reik W, Constancia M, Dean W, et al. Igf2 imprinting in development and disease. *Int J Dev Biol.* 2000;44: 145-150.
[72] Engel N, Bartolomei MS. Mechanisms of insulator function in gene regulation and genomic imprinting. *Int Rev Cytol.* 2003;232: 89-127.
[73] Murrell A, Heeson S, Reik W. Interaction between differentially methylated regions partitions the imprinted genes Igf2 and H19 into parent-specific chromatin loops. *Nat Genet.* 2004;36: 889-893.
[74] Sullivan MJ, Taniguchi T, Jhee A, Kerr N, Reeve AE. Relaxation of IGF2 imprinting in Wilms tumours associated with specific changes in IGF2 methylation. *Oncogene.* 1999;18: 7527-7534.
[75] Zhang H, Niu B, Hu JF, et al. Interruption of intrachromosomal looping by CTCF decoy proteins abrogates genomic imprinting of human insulin-like growth factor II. *J Cell Biol.* 2011;193: 475-487.
[76] Vu TH, Nguyen AH, Hoffman AR. Loss of IGF2 imprinting is associated with abrogation of long-range intrachromosomal interactions in human cancer cells. *Hum Mol Genet.* 2010;19: 901-919.
[77] Qiu X, Vu TH, Lu Q, et al. A complex deoxyribonucleic Acid looping configuration associated with the silencing of the maternal igf2 allele. *Mol Endocrinol.* 2008;22: 1476-1488.
[78] Kurukuti S, Tiwari VK, Tavoosidana G, et al. CTCF binding at the H19 imprinting control region mediates maternally inherited higher-order chromatin conformation to restrict enhancer access to Igf2. *Proc Natl Acad Sci U S A.* 2006;103: 10684-10689.
[79] Yoon YS, Jeong S, Rong Q, Park KY, Chung JH, Pfeifer K. Analysis of the H19ICR insulator. *Mol Cell Biol.* 2007;27: 3499-34510.
[80] Ling JQ, Li T, Hu JF, et al. CTCF mediates interchromosomal colocalization between Igf2/H19 and Wsb1/Nf1. *Science.* 2006;312: 269-272.
[81] Cao R, Wang L, Wang H, et al. Role of histone H3 lysine 27 methylation in Polycomb-group silencing. *Science.* 2002;298: 1039-1043.

[82] Czermin B, Melfi R, McCabe D, Seitz V, Imhof A, Pirrotta V. Drosophila enhancer of Zeste/ESC complexes have a histone H3 methyltransferase activity that marks chromosomal Polycomb sites. *Cell.* 2002;111: 185-196.
[83] Kirmizis A, Bartley SM, Kuzmichev A, et al. Silencing of human polycomb target genes is associated with methylation of histone H3 Lys 27. *Genes Dev.* 2004;18: 1592-1605.
[84] Rajasekhar VK, Begemann M. Concise review: roles of polycomb group proteins in development and disease: a stem cell perspective. *Stem Cells.* 2007;25: 2498-2510.
[85] Jaenisch R, Young R. Stem cells, the molecular circuitry of pluripotency and nuclear reprogramming. *Cell.* 2008;132: 567-582.
[86] Sparmann A, van Lohuizen M. Polycomb silencers control cell fate, development and cancer. *Nat Rev Cancer.* 2006;6: 846-856.
[87] Jones PA, Baylin SB. The epigenomics of cancer. *Cell.* 2007;128: 683-692.
[88] Payer B, Lee JT. X chromosome dosage compensation: how mammals keep the balance. *Annu Rev Genet.* 2008;42: 733-772.
[89] Wu HA, Bernstein E. Partners in imprinting: noncoding RNA and polycomb group proteins. *Dev Cell.* 2008;15: 637-638.
[90] Pien S, Grossniklaus U. Polycomb group and trithorax group proteins in Arabidopsis. *Biochim Biophys Acta.* 2007;1769: 375-382.
[91] Whitcomb SJ, Basu A, Allis CD, Bernstein E. Polycomb Group proteins: an evolutionary perspective. *Trends Genet.* 2007;23: 494-502.
[92] Li T, Hu JF, Qiu X, et al. CTCF regulates allelic expression of Igf2 by orchestrating a promoter-polycomb repressive complex 2 intrachromosomal loop. *Mol Cell Biol.* 2008;28: 6473-6482.
[93] Wang H, Ge S, Qian G, et al. Restoration of IGF2 imprinting by polycomb repressive complex 2 docking factor SUZ12 in colon cancer cells. *Exp Cell Res.* 2015;338: 214-221.
[94] Pan Y, He B, Li T, et al. Targeted tumor gene therapy based on loss of IGF2 imprinting. *Cancer Biol Ther.* 2010;10: 290-298.
[95] Collier RJ. Diphtheria toxin: mode of action and structure. *Bacteriol Rev.* 1975;39: 54-85.
[96] Li Y, McCadden J, Ferrer F, et al. Prostate-specific expression of the diphtheria toxin A chain (DT-A): studies of inducibility and specificity of expression of prostate-specific antigen promoter-driven DT-A adenoviral-mediated gene transfer. *Cancer Res.* 2002;62: 2576-2582.

[97] Hall PD, Virella G, Willoughby T, Atchley DH, Kreitman RJ, Frankel AE. Antibody response to DT-GM, a novel fusion toxin consisting of a truncated diphtheria toxin (DT) linked to human granulocyte-macrophage colony stimulating factor (GM), during a phase I trial of patients with relapsed or refractory acute myeloid leukemia. *Clin Immunol.* 2001;100: 191-197.

[98] Pan Y, He B, Lirong Z, et al. Gene therapy for cancer through adenovirus vectormediated expression of the Ad5 early region gene 1A based on loss of IGF2 imprinting. *Oncol Rep.* 2013;30: 1814-1822.

[99] Nevins JR. Mechanism of activation of early viral transcription by the adenovirus E1A gene product. Cell. 1981;26: 213-220.

[100] Everts B, van der Poel HG. Replication-selective oncolytic viruses in the treatment of cancer. *Cancer Gene Ther.* 2005;12: 141-161.

[101] Nie ZL, Pan YQ, He BS, et al. Gene therapy for colorectal cancer by an oncolytic adenovirus that targets loss of the insulin-like growth factor 2 imprinting system. *Mol Cancer.* 2012;11: 86.

BIBLIOGRAPHY

Applied plant genomics and biotechnology
LCCN   2014948687
Type of material   Book
Main title   Applied plant genomics and biotechnology/edited by Palmiro Poltronieri, Yiguo Hong.
Published/Produced   Amsterdam; Boston; Elsevier/WP, Woodhead Publishing is an imprint of Elsevier, [2015] ©2015
Description   xxxviii, 315 pages; illustrations; 24 cm.
Links   Publisher description http://www.loc.gov/catdir/enhancements/fy1606/2014948687-d.html
ISBN   9780081000687
                        0081000685
LC classification   QK981 .A67 2015
Related names   Poltronieri, Palmiro, editor.
                        Hong, Yiguo, editor.
Contents   Transgenic, cisgenic and novel plant products: challenges in regulation and safety assessment/Palmiro Poltronieri and Ida Barbara Reca -- What turns on and off the cytokinin metabolisms and beyond/Eva Jiskrová, Ivona Kubalová and Yoshihisa Ikeda -- Apple allergens genomics and biotechnology: unravelling the determinanats of apple allergenicity/Federica Savazzini, Giampaolo Ricci and Stefano Tartarini -- Non-food interventions: exploring plant biotechnology applications to therapeutic protein production/Matteo Busconi, Mariangela Marudelli and Corrado Fogher -- In planta produced virus-like

particles as candidate vaccines/Slavica Matić and Emanuela Noris -- Biotechnology of euphorbiaceae (Jatropha curcas, Manihot esculenta, Ricinus communis)/Fatemah Maghuly, Johann Vollmann and Margit Laimer -- Regulation framework for flowering/Tiziana Sgamma and Stephen Jackson -- Epigenetic regulation during fleshy fruit development and ripening/Emeline Teyssier, Lisa Boureauv, Weiwei Chen, Ruie Lui, Charlotte Degraeve-Guibault, Linda Stammitti, Yiguo Hong and Philippe Gallusci -- Tomato fruit quality improvement facing the functional genomics revolution/Dominique Rolin, Emeline Teyssier, Yiguo Hong and Philippe Gallusci -- Rice genomics and biotechnology/Dawei Xue, Hua Jiang and Qian Qian -- Genome-wide DNA methylation in tomato/Rupert Fray and Silin Zhong -- Recent application of biotechniques for the improvement of mango research/Mohammad Sorof Uddin and Qi Cheng -- Cotton genomics and biotechnology/Hao Juan and Sun Yuqiang -- Virus technology for functional genomics in plants/Cheng Qin, Qi Zhang, Meiling He, Junhua Kong, Bin LI, Atef Mohamed, Weiwei Chen, Pengcheng Zhang, Xian Zhang, Zhiming Yu, Tongfei Lai, Nongnong Shi, Toba Osman and Yiguo Hong -- PARP proteins, NAD, epigenetics, antioxidative response to abiotic stress/Palmiro Poltronieri and Masanao Miwa -- Applied oilseed rape marker technology and genomics/Christian Obermeier and Wolfgang Friedt.

Subjects Plant genetics.
Genomics.
Biotechnology.
Biotechnology.
Genomics.
Plant genetics.

Notes Includes bibliographical references and index.
Series Woodhead Publishing series in biomedicine; number 72
Woodhead Publishing series in biomedicine; no. 72.

Bacterial genomics; genome organization and gene expression tools

LCCN	2014027746
Type of material	Book
Personal name	Seshasayee, Aswin Sai Narain, author.
Main title	Bacterial genomics; genome organization and gene expression tools/Aswin Sai Narain Seshasayee.
Published/Produced	Delhi, India; New York; Cambridge University Press, 2015.
Description	xi, 211 pages; color illustrations; 24 cm.
ISBN	9781107079830 (hardback)
LC classification	QH434 .S45 2015
Summary	"Discusses the application of genomic tools in the study of bacterial adaptation and provides review of recent research in the field of bacterial research"-- Provided by publisher.
Contents	Introduction; bacterial genomes and gene expression -- Comparative genomics in the era of Sanger sequencing -- Studying bacterial genome variation with microarrays -- Studying bacterial genomes using next-generation sequencing -- Genome-scale analysis of gene expression and its regulation in bacteria -- DNA methylation in bacteria; a case for bacterial epigenetics.
Subjects	Bacteria--genetics.
	Genome, Bacterial--genetics.
	Gene Expression Regulation, Bacterial.
	Gene Expression.
	Models, Genetic.
Notes	Includes bibliographical references and index.

Bioinformatics for high throughput sequencing

LCCN	2011937571
Type of material	Book
Main title	Bioinformatics for high throughput sequencing/Naiara Rodríguez-Ezpeleta, Michael Hackenberg, Ana M. Aransay, editors.
Published/Created	New York, NY; Springer, c2012.
Description	xii, 255 p.; ill. (some col.); 24 cm.
Links	Table of contents only http://www.loc.gov/catdir/ enhancements/fy1208/2011937571-t.html

ISBN	Publisher description http://www.loc.gov/catdir/enhancements/fy1208/2011937571-d.html 9781461407812 (alk paper) 1461407818 (alk. paper)
LC classification	QH441.2 .B563 2012
Related names	Rodríguez-Ezpeleta, Naiara. Hackenberg, Michael, 1972- Aransay, Ana M.
Contents	Overview of sequencing technology platforms/Samuel Myllykangas, Jason Buenrostro, and Hanlee P. Ji -- Applications of high-throughput sequencing/Rodrigo Goya, Irmtraud M. Meyer, and Marco A. Marra -- Computational infrastructure and basic data analysis for high-throughput sequencing/David Sexton -- Base-calling for bioinformaticians/Mona A. Sheikh and Yaniv Erlich -- De novo short-read assembly/Douglas W. Bryant Jr. and Todd C. Mockler -- Short-read mapping/Paolo Ribeca -- DNA-protein interaction analysis (ChIP-Seq)/Geetu Tuteja -- Generation and analysis of genome-wide DNA methylation maps/Martin Kerick, Axel Fischer, and Michal-Ruth Schweiger -- Differential expression for RNA sequencing (RNA-Seq) data; mapping, summarization, statistical analysis, and experimental design/Matthew D. Young ... [et al.] -- MicroRNA expression profiling and discovery/Michael Hackenberg -- Dissecting splicing regulatory network by integrative analysis of CLIP-Seq data/Michael Q. Zhang -- Analysis of metagenomics data/Elizabeth M. Glass and Folker Meyer -- High-throughput sequencing data analysis software; current state and future developments/Konrad Paszkiewicz and David J. Studholme.
Subjects	Sequence alignment (Bioinformatics) Genomics--Data processing.
Notes	Includes bibliographical references and index.

Cancer epigenetics; biomolecular therapeutics for human cancer

LCCN	2010041037
Type of material	Book
Main title	Cancer epigenetics; biomolecular therapeutics for human cancer/edited by Antonio Giordano, Marcella Macaluso.

Published/Created	Hoboken, N.J.; Wiley-Blackwell, c2011.
Description	xv, 390 p.; ill.; 24 cm.
ISBN	9780471710967 (cloth)
LC classification	RC268.4 C3492 2011
Related names	Giordano, Antonio, MD.
	Macaluso, Marcella.
Summary	"Biomolecular Therapeutics for Human Cancer is the only resource to focus on biomolecular approaches to cancer therapy. Its presentation of the latest research in cancer biology reflects the interdisciplinary nature of the field and aims to facilitate collaboration between the basic, translational, and clinical sciences"--Provided by publisher.
Subjects	Cancer--Genetic aspects.
	Cancer--Treatment.
	Epigenesis.
	DNA--Methylation.
	Neoplasms--genetics.
	Neoplasms--therapy.
	DNA Methylation.
	Drug Design.
	Epigenesis, Genetic.
	Histones--metabolism.
Notes	Includes bibliographical references and index.

Cancer gene profiling; methods and protocols

LCCN	2009930638
Type of material	Book
Main title	Cancer gene profiling; methods and protocols/edited by Robert Grützmann and Christian Pilarsky.
Published/Created	New York; Humana Press, c2010.
Description	xi, 450 p.; ill.; 28 cm.
ISBN	9781934115763 (alk. paper)
	1934115762 (alk. paper)
	9781597455459 (ebook)
LC classification	RC268.4 .C3496 2010
Related names	Grützmann, Robert.
	Pilarsky, Christian.
Contents	Organizational issues in providing high-quality human tissues and clinical information for the support of biomedical research/Walter C. Bell, Katherine C. Sexton,

and William E. Grizzle -- Manual microdissection/Glen Kristiansen -- Laser microdissection/Anja Rabien -- Tissue microarrays/Ana-Maria Dancau ... [et al.] -- A decade of cancer gene profiling; from molecular portraits to molecular function/Henri Sara, Olli Kallioniemi, and Matthias Nees -- Mining expressed sequence tag (EST) libraries for cancer-associated genes/Armin O. Schmitt -- Automated fluorescent differential display for cancer gene profiling/Jonathan D. Meade ... [et al.] -- Manual microdissection combined with antisense RNA-longSAGE for the analysis of limited cell numbers/Jutta Lüttges, Stephan A. Hahn, and Anna M. Heidenblut -- Quantitative DNA methylation profiling on a high-density oligonucleotide microarray/Anne Fassbender ... [et al.] -- Single-nucleotide polymorphism (SNP) analysis to associate cancer risk/Julie Earl and William Greenhalf -- Application of proteomics in cancer gene profiling; two-dimensional difference in gel electrophoresis (2D-DIGE)/Deepak Hariharan, Mark E. Weeks, and Tatjana Crnogorac-Jurcevic -- Search for and identification of novel tumor-associated autoantigens/Karsten Conrad ... [et al.] -- Integrative oncogenomic analysis of microarray data in hematologic malignancies/Jose A. Martínez-Climent ... [et al.] -- Cancer gene profiling in pancreatic cancer/Felip Vilardell and Christine A. Iacobuzio-Donahue -- Cancer gene profiling in prostate cancer/Adam Foye and Phillip G. Febbo -- Cancer gene profiling for response prediction/B. Michael Ghadimi and Marian Grade -- The EGFR pathway as an example for genotype; phenotype correlation in tumor genes/Ulrike Mogck, Eray Goekkurt, and Jan Stoehlmacher -- Quantitation of CD39 gene expression in pancreatic tissue by real-time polymerase chain reaction/Martin Loos, Beat Künzli, nad Helmut Friess -- Functional profiling methods in cancer/Joaquín Dopazo -- Calibration of microarray gene-expression data/Hans Binder, Stephan Preibisch, and Hilmar Berger -- Meta-analysis of cancer gene-profiling data/Xinan Yang and Xiao Sun -- Target gene discovery for novel therapeutic agents in cancer treatment/Ole Ammerpohl, Sanjay Tiwari, and Holger Kalthoff.

Subjects	Cancer genes.
	DNA microarrays.
	Neoplasms--genetics--Laboratory Manuals.
	Gene Expression Profiling--methods--Laboratory Manuals.
	Microarray Analysis--methods--Laboratory Manuals.
Notes	Includes bibliographical references and index.
Series	Springer protocols
	Methods in molecular biology, 1064-3745; 576
	Springer protocols.
	Methods in molecular biology (Clifton, N.J.); v. 576.

Cervical cancer; methods and protocols

LCCN	2014952732
Type of material	Book
Main title	Cervical cancer; methods and protocols/edited by Daniel Keppler, Athena W. Lin.
Published/Created	New York; Humana Press, c2015.
Description	xiv, 413 p.; ill. (some col.); 27 cm.
ISBN	9781493920129 (hbk.; acid-free paper)
	149392012X (hbk.; acid-free paper)
LC classification	RC280.U8 C4722 2015
Related names	Keppler, Daniel, 1957- edt
	Lin, Athena W., 1962- edt
	Springer Science+Business Media, copyright holder.
Summary	Representing the most relevant procedures and technologies aiding the advance of the field of HPV-mediated carcinogenesis of the cervix and other anatomical regions of squamocolumnar transition, such as the anorectum, penis, and oropharynx, Cervical Cancer: Methods and Protocols compiles a detailed collection of practical chapters. The first half of the book covers HPV types, pathogenesis of cervical cancer (CxCA), prevention, and novel potential drug targets, while the second half explores pathology, genomics, modeling of CxCA, and experimental therapeutic strategies. Written in the highly successful Methods in Molecular Biology series format, chapters include introductions to their respective topics, lists of the necessary materials and reagents, step-by-step, readily reproducible laboratory

Contents protocols, and tips on troubleshooting and avoiding known pitfalls. Authoritative and vital, Cervical Cancer: Methods and Protocols serves as a valuable resource to both bench scientists and clinicians who step into the realm of high-risk HPVs and CxCA for the first time or those who wish to learn novel approaches or expand their toolbox for the study of CxCA.

Evolution and classification of oncogenic human papillomavirus types and variants associated with cervical cancer/Zigui Chen, Luciana Bueno de Freitas, and Robert D. Burk -- Real-time PCR approach based on SPF10 primers and the INNO-LiPA HPV genotyping extra assay for the detection and typing of human papillomavirus/M. Isabel Micalessi, Gaëlle A. Boulet, and Johannes Bogers -- Replication of human papillomavirus in culture/Eric J. Ryndock, Jennifer Biryukov, and Craig Meyers -- HPV binding assay to laminin-332/integrin [alpha]6[beta]4 on human keratinocytes/Sarah A. Brendle and Neil D. Christensen -- Methods to assess the nucleocytoplasmic shuttling of the HPV E1 helicase and its effects on cellular proliferation and induction of a DNA damage response/Michaël Lehoux, Amélie Fradet-Turcotte, and Jacques Archambault -- Genetic methods for studying the role of viral oncogenes in the HPV life cycle/Jason M. Bodily -- Robust HPV-18 production in organotypic cultures of primary human keratinocytes/Hsu-Kun Wang, Thomas R. Broker, and Louise T. Chow -- High-throughput cellular assay to quantify the p53- degradation activity of E6 from different human papillomavirus types/David Gagnon and Jacques Archambault -- Retroviral expression of human cystatin genes in HeLa cells/Crystal M. Diep, Gagandeep Kaur, Daniel Keppler, and Athena W. Lin -- Molecular analysis of human papillomavirus virus-like particle activated langerhans cells in vitro/Andrew W. Woodham, Adam W. Woodham, Adam B. Raff, Diane M. Da Silva, and W. Martin Kast -- Selective silencing of gene target expression by siRNA expression plasmids in human cervical cancer cells/Oscar Peralta-Zaragoza [and eleven others] -- Silencing of E6/E7 expression in cervical cancer stem- like cells/Wenyi Gu, Nigel McMillan, and Chengzhong Yu -- Two-step procedure for evaluating experimentally induced DNA damage; texas red and comet assays/Gina J. Ferris, Lauren B. Shunkwiler, and Charles A. Kunos -- Measurement of deubiquitinating enzyme Activity via a suicidal HA-Ub-VS probe/Colleen Rivard and Martina Bazzaro -- Immunocytochemical analysis of the cervical pap smear/Terry K. Morgan and Michelle Berlin -- Diagnosis of HPV-negative, gastric-type adenocarcinoma of the endocervix/Edyta C. Pirog -- Targeting of the HPV-16 E7 protein by RNA aptamers/Julia Dolores Toscano-Garibay, María Luisa Benítez-Hess, and Luis Marat Alvarez-Salas -- Use of MYBL2 as a novel candidate biomarker of cervical cancer/Cara M. Martin [and others] -- Fixation methods for the preservation of morphology, RNAs, and proteins in paraffin-Embedded human cervical cancer cell xenografts in mice/Yoko Matsuda and Toshiyuki Ishiwata -- Assessment of the HPV DNA methylation status in cervical lesions/Mina Kalantari and Hans-Ulrich Bernard -- MeDIP-on-chip for methylation profiling/Yaw-Wen Hsu, Rui-Lan Huang, and Hung-Cheng Lai -- Use of DBD-FISH for the study of cervical cancer progression/Elva I. Cortés-Gutiérrez [and four others] -- Quantitative and high-throughput assay of human papillomavirus DNA replication/David Gagnon, Amélie Fradet-Turcotte, and Jacques Archambault -- Native human papillomavirus production, quantification, and infectivity analysis/Jennifer Biryukov, Linda Cruz, Eric J. Ryndock, and Craig Meyers -- Functional analysis of HPV-like particle-activated langerhans cells in vitro/Lisa Yan, Andrew W. Woodham, Diane M. Da Silva, and W. Martin Kast -- Assessment of the radiation sensitivity of cervical cancer cell lines/Mayil Vahanan Bose and Thangarajan Rajkumar -- Mouse model of cervicovaginal papillomavirus infection/Nicolas Çuburu, Rebecca J. Cevio, Cynthia

	D. Thompson, and Patricia M. Day -- Establishment of orthotopic primary cervix cancer xenografts/Naz Chaudary, Karolina Jaluba, Melania Pintilie, and Richard P. Hill -- Generation of K14-E7/[delta]N87[beta] cat double transgenic mice as a model of cervical cancer/Gülay Bulut and Aykut Üren.
Subjects	Cervix uteri--Cancer.
	Uterine Cervical Neoplasms--Laboratory Manuals.
	Human Papillomavirus DNA Tests--Laboratory Manuals--methods.
	Cervix uteri--Cancer.
Form/Genre	Laboratory Manuals.
Notes	Includes bibliographical references and index.
Series	Methods in molecular biology, 1064-3745; 1249
	Springer protocols, 1949-2448
	Methods in molecular biology (Clifton, N.J.); v. 1249, 1064-3745
	Springer protocols (Series), 1949-2448

Chromatin remodeling; methods and protocols

LCCN	2011943306
Type of material	Book
Main title	Chromatin remodeling; methods and protocols/edited by Randall H. Morse.
Published/Created	New York; Humana Press, c2012.
Description	xii, 453 p.; ill. (some col.); 27 cm.
ISBN	9781617794766 (hbk.; alk. paper)
	1617794767 (hbk.; alk. paper)
LC classification	QH599 .C457 2012
Related names	Morse, Randall H.
Contents	Strain construction and screening methods for a yeast histone H3/H4 mutant library/Junbiao Dai and Jef D. Boeke -- Measuring dynamic changes in histone modifications and nucleosome density during activated transcription in budding yeast/Chhabi K. Govind, Daniel Ginsburg, and Alan G. Hinnebusch -- Monitoring the effects of chromatin remodelers on long-range interactions in vivo/Christine M. Kiefer and Ann Dean -- Measuring nucleosome occupancy in vivo by micrococcal nuclease/Gene O. Bryant --

Analysis of nucleosome positioning using a nucleosome-scanning assay/Juan Jose Infante, G. Lynn Law, and Elton T. Young -- Assaying chromatin structure and remodeling by restriction enzyme accessibility/Kevin W. Trotter and Trevor K. Archer -- Generation of DNA circles in yeast by inducible site-specific recombination/Marc R. Gartenberg -- An efficient purification system for native minichromosome from Saccharomyces cerevisiae/Ashwin Unnikrishnan ... [et al.] -- Simultaneous single-molecule detection of endogenous C-5 DNA methylation and chromatin accessibility using MAPit/Russell P. Darst ... [et al.] -- Analysis of stable and transient protein-protein interactions/Stephanie Byrum ... [et al.] -- Monitoring dynamic binding of chromatin proteins in vivo by fluorescence recovery after photobleaching/Florian Mueller ... [et al.] -- Monitoring dynamic binding of chromatin proteins in vivo by fluorescence correlation spectroscopy and temporal image correlation spectroscopy/Davide Mazza ... [et al.] -- Analysis of chromatin structure in plant cells/Mala Singh ... [et al.] -- Analysis of histones and histone variants in plants/Ila Trivedi ... [et al.] -- Reconstitution of modified chromatin templates for in vitro functional assays/Miyong Yun ... [et al.] -- A defined in vitro system to study ATP-dependent remodeling of short chromatin fibers/Verena K. Maier and Peter B. Becker -- In vitro reconstitution of in vivo-like nucleosome positioning on yeast DNA/Christian J. Wippo and Philipp Korber -- Activator-dependent acetylation of chromatin model systems/Heather J. Szerlong and Jeffrey C. Hansen -- Mapping assembly favored and remodeled nucleosome positions on polynucleosomal templates/Hillel I. Sims, Chuong D. Pham, and Gavin R. Schnitzler -- Analysis of changes in nucleosome conformation using fluorescence resonance energy transfer/Tina Shahian and Geeta J. Narlikar -- Preparation of nucleosomes containing a specific H2A-H2A cross-link forming a DNA-constraining loop structure/Ning Liu and Jeffrey J. Hayes --

	Sulfyhydryl-reactive site-directed cross-linking as a method for probing the tetrameric structure of histones H3 and H4 -/ Andrew Bowman and Tom Owen-Hughes -- Genomic approaches for determining nucleosome occupancy in yeast Kyle Tsui ... [et al.] -- Genome-wide approaches to determining nucleosome occupancy in metazoans using MNase-Seq/Kairong Cui and Keji Zhao -- Salt fractionation of nucleosomes for genome-wide profiling/Sheila S. Teves and Steven Henikoff -- Quantitative analysis of genome-wide chromatin remodeling/Songjoon Baek, Myong-Hee Sung, and Gordon L. Hager -- Computational analysis of nucleosome positioning/Itay Tirosh.
Subjects	Chromatin--Structure.
	Chromatin--Laboratory manuals.
	Chromatin Assembly and Disassembly--Laboratory Manuals.
Notes	Includes bibliographical references and index.
Series	Methods in molecular biology, 1940-6029; 833
	Springer protocols
	Methods in molecular biology (Clifton, N.J.); v. 833.
	Springer protocols.

Developmental toxicology; methods and protocols
LCCN	2012937857
Type of material	Book
Main title	Developmental toxicology; methods and protocols/edited by Craig Harris, Jason M. Hansen.
Published/Created	New York; Humana Press; Springer, c2012.
Description	xii, 471 p.; ill. (some col.); 27 cm.
ISBN	9781617798665 (alk. paper)
	1617798665 (alk. paper)
	9781617798672 (e-ISBN)
	1617798673 (e-ISBN)
LC classification	RA1224.45 .D485 2012
	QH506 .M45 v.889
Related names	Harris, Craig, Ph.D.
	Hansen, Jason M., Ph.D.
Contents	In vivo models of developmental toxicology/Jason M. Hansen -- Caenorhabditis elegans as a model in

developmental toxicology/Windy A. Boyd, Marjolein V. Smith, and Jonathan H. Freedman -- Zebrafish embryo developmental toxicology assay/Julieta M. Panzica-Kelly, Cindy X. Zhang, and Karen Augustine-Rauch -- Gene knockdown by morpholino-modified oligonucleotides in the Zebrafish (Danio rerio) model; applications for developmental toxicology/Alicia R. Timme-Laragy, Sibel I. Karchner, and Mark E. Hahn -- Amphibian model for studies of developmental reproductive toxicity/Cecilia Berg -- Avian models in teratology and developmental toxicology/Susan M. Smith, George R. Flentke, and Ana Garic -- Overview of in vitro models in developmental toxicology/Craig Harris -- Primary cell and micromass culture in assessing developmental toxicity/M. Pratten -- Embryonic stem cell test; stem cell use in predicting developmental cardiotoxicity and osteotoxicity/Beatrice Kuske, Polina Y. Pulyanina, and Nicole I. zur Nieden -- Mouse embryonic stem cell adherent cell differentiation and cytotoxicity assay/Marianne Barrier ... [et al.] -- Murine limb bud in culture as an in vitro teratogenicity test System/France-Helene Paradis, Chunwei Huang, and Barbara F. Hales -- Rodent whole embryo culture/Craig Harris -- Rabbit whole embryo culture/Valerie A. Marshall and Edward W. Carney -- Assessment of xenobiotic biotransformation including reactive oxygen species generation in the embryo using benzene as an example/Helen J. Renaud, Allison Rutter, and Louise M. Winn -- Methodological approaches to cytochrome P450 profiling in embryos/Jared V. Goldstone and John J. Stegeman -- Analysis of Nrf2-mediated transcriptional induction of antioxidant response in early embryos/Shao-yu Chen -- Regulation and control of AP-1 binding activity in embryotoxicity/Terence R.S. Ozolins -- Thioredoxin redox status assessment during embryonic development; the redox western/Jason M. Hansen -- Methods for the determination of plasma or tissue glutathione levels/Trent E. Tipple and Lynette K. Rogers -- Oxidative stress, thiols, and redox

	profiles/Craig Harris and Jason M. Hansen -- Review of toxicogenomic approaches in developmental toxicology/Joshua F. Robinson, Jeroen L.A. Pennings, and Aldert H. Piersma -- Epigenetic approaches and methods in developmental toxicology: role of HDAC inhibition in teratogenic events/Elena Menegola, Graziella Cappelletti, and Francesca Di Renzo -- DNA methylation screening and analysis/Karilyn E. Sant, Muna S. Nahar, and Dana C. Dolinoy -- Assessment of histiotrophic nutrition using fluorescent probes/Jeffrey Ambroso and Craig Harris -- Diabetic embryopathy/Ulf J. Eriksson and Parri Wentzel -- Gene expression analysis in developing embryos; in situ hybridization/Siew-Ging Gong -- Assessment of gross fetal malformations; the modernized wilson technique and skeletal staining/Robert E. Seegmiller ... [et al.].
Subjects	Developmental toxicology--Laboratory manuals. Abnormalities, Drug-Induced--Laboratory Manuals. Embryonic Development--drug effects--Laboratory Manuals. Fetal Development--drug effects--Laboratory Manuals. Toxicology--methods--Laboratory Manuals.
Notes	Includes bibliographical references and index.
Series	Methods in molecular biology, 1064-3745; 889 Springer Protocols Methods in molecular biology (Clifton, N.J.); v. 889. 1064-3745 Springer protocols.

DNA methylation; principles, mechanisms and challenges

LCCN	2012949964
Type of material	Book
Main title	DNA methylation; principles, mechanisms and challenges/Tatiana V. Tatarinova and Gaurav Sablok, editors.
Published/Produced	New York; Nova Science, [2013] ©2013
Description	184 pages; illustrations; 26 cm.
ISBN	1624171281

	9781624171284
LC classification	QP624.5.M46 D633 2013
Related names	Tatarinova, Tatiana V., editor.
	Sablok, Gaurav, editor.
Subjects	DNA--Methylation.
	DNA Methylation--genetics.
	DNA--Methylation.
Notes	Includes bibliographical references and index.
Series	Genetics - research and issues
	Genetics--research and issues series.

Engineered zinc finger proteins; methods and protocols

LCCN	2010930513
Type of material	Book
Main title	Engineered zinc finger proteins; methods and protocols/edited by Joel P. Mackay, David J. Segal.
Published/Created	New York; Humana Press, c2010.
Description	xiv, 500 p.; ill.; 26 cm.
ISBN	9781607617525 (alk. paper)
	1607617528 (alk. paper)
	9781607617532 (e-ISBN)
	1607617536 (e-ISBN)
LC classification	QP552.Z55 E54 2010
Related names	Mackay, Joel P.
	Segal, David J. (David Jay), 1966-
Contents	The generation of zinc finger proteins by modular assembly/Mital S. Bhakta and David J. Segal -- Engineering single Cys2His2 zinc finger domains using a bacterial cell-based two-hybrid selection system/Stacey Thibodeau-Beganny, Morgan L. Maeder, and J. Keith Joung -- Bipartite selection of zinc fingers by phage display for any 9-bp DNA target site/Jia-Ching Shieh -- Structure-based DNA-binding prediction and design/Andreu Alibés, Luis Serrano, and Alejandro D. Nadra -- Generation of cell-permeable artificial zinc finger protein variants/Takashi Sera -- Inhibition of viral transcription using designed zinc finger proteins/Kimberley A. Hoeksema and D. Lorne J. Tyrrell -- Modulation of gene expression using zinc finger-based artificial transcription factors/Sabine Stolzenburg ... [et al.] -- Construction of combinatorial libraries that encode

zinc finger-based transcription factors/Seokjoong Kim, Eun Ji Kim, and Jin-Soo Kim -- Silencing of gene expression by targeted DNA methylation; concepts and approaches/Renata Z. Jurkowska and Albert Jeltsch -- Remodeling genomes with artificial transcription factors (ATFs)/Adriana S. Beltran and Pilar Blancafort -- Transgenic mice expressing an artificial zinc finger regulator targeting an endogenous gene/Claudio Passananti ... [et al.] -- Artificial zinc finger nucleases for DNA cloning/Vardit Zeevi, Andriy Tovkach, and Tzvi Tzfira -- In vitro assessment of zinc finger nuclease activity/Toni Cathomen and Cem Şöllü -- Quantification of zinc finger nuclease-associated toxicity/Tatjana I. Cornu and Toni Cathomen -- A rapid and general assay for monitoring endogenous gene modification/Dmitry Y. Guschin ... [et al.] -- Engineered zinc finger proteins for manipulation of the human mitochondrial genome/Michal Minczuk -- High-efficiency gene targeting in Drosophila with zinc finger nucleases/Dana Carroll, Kelly J. Beumer, and Jonathan K. Trautman -- Using zinc finger nucleases for efficient and heritable gene disruption in zebrafish/Jasmine M. McCammon and Sharon L. Amacher -- A transient assay for monitoring zinc finger nuclease activity at endogenous plant gene targets/Justin P. Hoshaw ... [et al.] -- Validation and expression of zinc finger nucleases in plant cells/Andriy Tovkach, Vardit Zeevi, and Tzvi Tzfira -- Non-FokI-based zinc finger nucleases/Miki Imanishi, Shigeru Negi, and Yukio Sugiura -- Designing and testing chimeric zinc finger transposases/Matthew H. Wilson and Alfred L. George -- Seeing genetic and epigenetic information without DNA denaturation using sequence-enabled reassembly (SEER)/Jason R. Porter ... [et al.] -- Zinc finger-mediated live cell imaging in Arabidopsis roots/Beatrice I. Lindhout, Tobias Meckel, and Bert J. van der Zaal -- Biophysical analysis of the interaction of toxic metal ions and oxidants with the zinc finger domain of XPA/Andrea Hartwig, Tanja Schwerdtle, and Wojciech Bal -- Preparation and zinc-binding properties of multi-fingered zinc-sensing domains/John H. Laity and Linda S. Feng -- Using ChIP-seq technology to

	identify targets of zinc finger transcription factors/Henriette O'Geen, Seth Frietze, and Peggy J. Farnham -- Crystallization of zinc finger proteins bound to DNA/Nancy C. Horton and Chad K. Park -- Beyond DNA; zinc finger domains as RNA-binding modules/Josep Font and Joel P. Mackay.
Subjects	Zinc-finger proteins.
	Zinc Fingers--Laboratory Manuals.
	Transcription Factors--Laboratory Manuals.
Notes	Includes bibliographical references and index.
Additional formats	Also available online.
	Online version: Engineered zinc finger proteins. [Totowa, N.J.]; Humana Press; London; Springer [distributor], 2010 9781607617532 (OCoLC)658083485
Series	Methods in molecular biology, 1064-3745; v. 649
	Springer Protocols
	Methods in molecular biology (Clifton, N.J.); v.649.
	1064-3745
	Springer protocols.

Epigenetic alterations in oncogenesis

LCCN	2012945005
Type of material	Book
Main title	Epigenetic alterations in oncogenesis/Adam R. Karpf, editor.
Published/Created	New York; Springer, c2013.
Description	xv, 348 p.; ill. (some col.); 24 cm.
ISBN	9781441999665 (alk. paper)
	1441999663 (alk. paper)
	9781441999672 (e-Book)
	1441999671 (e-Book)
LC classification	RC268.5 .E65 2013
	R850.A1 A39 v.754 2013
Related names	Karpf, Adam R., 1969-
Contents	Epigenetic marks and mechanisms -- DNA methyltransferases, DNA damage repair, and cancer -- DNA hypomethylation and hemimethylation in cancer -- Ten eleven translocation enzymes and 5-hydroxymethylation in mammalian development and cancer -- Altered histone modifications in cancer -- Nucleosome occupancy and gene regulation during

	tumorigenesis -- Impact of epigenetic alterations on cancer biology -- Epigenetic regulation of miRNAs in cancer -- DNA hypomethylation and activation of germlike-specific genes in cancer -- APC and DNA demethylation in cell fate specification and intestinal cancer -- Epigenetic changes during cell transformation -- Epigenetic reprogramming of mesenchymal stem cells -- Clinical implications and analysis in methods -- Environmental toxicants, epigenetics, and cancer -- Blood-derived DNA methylation markers of cancer risk -- Epigenetic therapies in MDS and AML -- Epigenetic targeting therapies to overcome chemotherapy resistance -- Methods for cancer epigenome analysis.
Subjects	Carcinogenesis--Molecular aspects.
	Epigenetics.
	Cell Transformation, Neoplastic.
	Epigenesis, Genetic.
	Neoplasms--genetics.
Notes	Includes bibliographical references and index.
Series	Advances in experimental medicine and biology, 0065-2598; v.754
	Advances in experimental medicine and biology; v.754. 0065-2598

Epigenetic aspects of chronic diseases

LCCN	2011921946
Type of material	Book
Main title	Epigenetic aspects of chronic diseases/Helmtrud I. Roach, Felix Bronner, Richard O.C. Oreffo (editors).
Published/Created	London; Springer, c2011.
Description	xvii, 236 p.; ill. (some col.); 27 cm.
ISBN	9781848826434 (alk. paper)
	1848826435 (alk. paper)
	9781848826441 (e-ISBN)
	1848826443 (e-ISBN)
LC classification	RB156 .E65 2011
Related names	Roach, Helmtrud I.
	Bronner, Felix
	Oreffo, Richard O. C.

Contents	Epigenetics and chronic diseases; an overview/Rebecca Smith, Jonathan Mill -- Techniques to study DNA methylation and histone modification/Ester Lara ... [et al.] -- Mechanisms of epigenetic gene silencing/Marie-Pierre Lambert, Zdenko Herceg -- Mechanisms of epigenetic gene activation in disease; dynamics of DNA methylation and demethylation/Thierry Granbe, Edio Eligio Lourenco -- The role of histone demethylases in disease/Paul Cloos -- Autoimmune diseases/Travis Hughes, Amr H. Sawalha -- Epigenetics of rheumatoid arthritis/Aleksander M. Grabiec, Paul P. Tak, Kris A. Reedquist -- DNA methylation changes in osteoarthritis/Helmtrud I. Roach -- Epigenetics and Type 2 diabetes/Charlotte Ling, Tina Ronn, Marloes Dekker Nitert -- Epigenetic regulation of asthma and allergic diseases/Andrew L. Durham, Ivan M. Adcock -- Epigenetics in psychiatry/Hamid Mostafavi-Abdolmaleky, Stephen J. Glatt, Ming T. Tsuang -- Epigenetics and late-onset Alzheimer's disease/Axel Schumacher, Syed Bihaqi, Nasser H. Zawia -- Epigenetic mechanisms in the developmental origins of adult disease/Keith M. Godfrey ... [et al.] -- Targeting histone deacetylases in chronic obstructive pulmonary disease/Peter J. Barnes -- Clinical trials of epigenetic modifiers in the treatment of myelodysplastic syndrome/Lauren C. Suarez, Steven D. Gore.
Subjects	Chronic diseases. Chronic diseases--Prevention. Epigenesis. Chronic Disease. Epigenesis, Genetic. DNA Methylation--genetics. Histones--metabolism. Chronic disease. DNA methylation--Genetics. Histones--Metabolism.
Notes	Includes bibliographical references and index.

Epigenetic epidemiology
LCCN 2011943620
Type of material Book

Main title	Epigenetic epidemiology/Karin B. Michels, editor.
Published/Created	Dordrecht; New York; Springer Verlag, c2012.
Description	xii, 446 p.; ill. (some col.); 24 cm.
Links	Publisher description http://www.loc.gov/catdir/enhancements/fy1317/2011943620-d.html
	Table of contents only http://www.loc.gov/catdir/enhancements/fy1317/2011943620-t.html
ISBN	9789400724945 (alk. paper)
	9400724942 (alk. paper)
LC classification	RA651 .E69 2012
Related names	Michels, Karin B.
Contents	Human epigenome -- Considerations in the design, conduct, and interpretation of studies in epigenetic epidemiology -- Laboratory methods in epigenetic epidemiology -- Biostatistical methods in epigenetic epidemiology -- Epigenome changes during development -- Role of epigenetics in the development origins of health and disease -- Epigenetics and assisted reproductive technology -- Imprinting disorders of early childhood -- Utility of twins for epigenetic analysis -- Age-related variation in DNA methylation -- Influence of environmental factors on the epigenome -- Epigenetic epidemiology of cancer -- Epigenetic epidemiology of infectious diseases -- Epigenetic epidemiology of inflammation and rhematoid arthritis -- Asthma epigenetics; emergnece of a new paradigm? -- Epigenetic epidemiology of autism and other neurodevelopmental disorders -- Epigenetic epidemiology psychiatric disorders -- Epigenetic epidemiology of type 1 diabetes -- Epigenetic epidemiology of obesity, type 2 diabetes -- Epigenetic epidemiology of atherosclerosis.
Subjects	Epidemiology.
	Medical genetics.
	Epidemiologic Methods.
	Epigenomics--methods.
Notes	Includes bibliographical references and index.

Epigenetics
LCCN   2013040242

Type of material	Book
Personal name	Armstrong, Lyle, author.
Main title	Epigenetics/Lyle Armstrong.
Published/Produced	New York, NY; Garland Science, [2014]
Description	xii, 306 pages; illustrations; 28 cm
ISBN	9780815365112 (pbk.; alk. paper)
LC classification	RB155.5 .A76 2014
Summary	"The concept of epigenetics has been known about since the 1940s, but it is only in the last 10 years that research has shown just how wide ranging its effects are. It is now a very widely-used term, but there is still a lot of confusion surrounding what it actually is and does. Epigenetics brings together the structure and machinery of epigenetic modification, how epigenetic modification controls cellular functions, and the evidence for the relationship between epigenetics and disease. It will therefore be an invaluable source of information about all aspects of this subject for undergraduate students, graduate students, and professionals alike. Topics included are the role of epigenetics in stem cells, the involvement of epigenetics with cancer, and the role of epigenetics in mental health. Key Features - Describes the two forms of epigenetic modification, DNA methylation and histone acetylation, and how they take place. - Section on how epigenetics controls cell function, including cellular differentiation and consequently the role of epigenetics in stem cells - Section on the role of epigenetics in disease only includes diseases where there is clear evidence of epigenetic involvement - Glossary explaining all the terms involved in epigenetics - Full-color figures"--Provided by publisher.
Subjects	Epigenomics.
	Epigenesis, Genetic.
	Genetic Predisposition to Disease.
Notes	Includes bibliographical references and index.

Epigenetics; a reference manual
LCCN	2012472998
Type of material	Book

Main title	Epigenetics; a reference manual/edited by Jeffrey M. Craig and Nicholas C. Wong.
Published/Created	Norfolk; Caister Academic Press, c2011.
Description	xii, 449 p., 1 leaf of plates; ill. (some col.); 26 cm.
ISBN	9781904455882
	1904455883
LC classification	QH450 .E6555 2011
Related names	Craig, Jeffrey.
	Wong, Nicholas C.
Contents	Early life environment, DNA methylation and behaviour/Moshe Szyf -- Concepts in histone acetyltransferase biology/Anne K. Voss and Tim Thomas -- Murine models of transgenerational epigenetic inheritance/Jennifer E. Cropley and Catherine M. Suter -- The molecular mechanisms of mammalian X inactivation/Marnie E. Blewitt and Linden J. Gearing -- Epigenetic memory in plants: polycomb-group regulation of responses to low temperature/Sandra N. Oliver and E. Jean Finnegan -- Centromeres and telomeres/Emma L. Northrop and Lee H. Wong -- DNA sequence contribution to nucleosome distribution/Justin A. Fincher and Jonathan H. Dennis -- Macrosatellite epigenetics/Brian P. Chadwick -- Histones: dosage and degradation/Rakesh Kumar Singh, Johanna Paik and Akash Gunjan -- The epigenetic basis of cell fate specification and reprogramming/Hongchang Cui -- DNA methylation changes in cancer/Samson Mani and Zdenko Herceg -- Variant Histones H2A and cancer development/Danny Rangasamy -- 5-Methlcytosine as a modification in RNA/Jeffrey E. Squires and Thomas Preiss -- Paramutation in plants/Mario A. Arteaga-Vazquez and Ana E. Dorantes-Acosta -- Lessons from DNMT3L-dependent methylation druing gametogenesis/Sarah A. Kinkel and Hamish S. Scott -- Non-coding RNA: an overview/Alka Saxena and Piero Carninci -- Bisulfite-enabled technologies/Miina Ollikainen -- Methylation-sensitive high resolution melting for the rapid analysis of DNA methylation/Thomas Mikeska and Alexander Dobrovic -- Microarray mapping of

	nucleosome position/Brian Spetman, Sarah Lueking, Brooke Roberts Druliner and Jonathan H. Dennis -- Enzymatic approaches for genome DNA methylation profiling/Benjamin Chanrion, Yurong Xin and Fatemeh Haghighi -- ChIP sequencing/Sebastian Lunke and Assam El-Osta -- Genome-wide DNA Methylation Analysis/Marcel W. Coolen and Susan J. Clark -- Bioinformatics analysis of epigenomic methylation patterns in the era of massively parallel sequencing/Mark D. Robinson, Bryan Beresford-Smith, Anthony Kaspi and Izhak Haviv -- Genetic resources for the study of epigenetic gene regulation in Maize/Andre Irsigler and Karen M. McGinnis -- Online resources and tools for epigeneticists/Nicholas C. Wong -- Educational resources for epigenetics/Yuk Jing Loke and Jeffrey M. Craig.
Subjects	Genetic regulation.
	Gene expression.
	Epigenesis, Genetic.
	Epigenomics.
	DNA Methylation.
Notes	Includes bibliographical references and index.

Epigenetics and development

LCCN	2013427822
Type of material	Book
Main title	Epigenetics and development/edited by Edith Heard.
Edition	1st ed.
Published/Created	San Diego, Calif.; Academic Press, 2013.
Description	xv, 334 p., [16] p. of plates; ill. (some col.); 24 cm.
Links	Publisher description http://www.loc.gov/catdir/enhancements/fy1606/2013427822-d.html
ISBN	9780124160279
	0124160271
LC classification	QH450 .E6558 2013
	QL951 .C8 v.104
Related names	Heard, Edith.
Contents	Mechanisms and dynamics of heterochromatin formation during mammalian development; closed paths and open questions/Anas Fadloun, André Eid,

	Maria-Elena Torres-Padilla -- Functions of DNA methylation and hydroxymethylation in mammalian development/Sylvain Guibert, Michael Weber -- Epigenetic marking of the zebrafish developmental program/Ingrid S. Andersen, Leif C. Lindeman, Andrew H. Reiner, Olga Østrup, Håvard Aanes, Peter Aleström, Philippe Collas -- Chromatin architecture and hox gene collinearity/Daan Noordermeer, Denis Duboule -- Primordial germ-cell development and epigenetic reprogramming in mammals/Harry G. Leitch, Walfred W.C. Tang, M. Azim Surani -- Epigenetics and development in plants; green light to convergent innovations/Daniel Grimanelli, François Roudier -- Reprogramming and the pluripotent stem cell cycle/Tomomi Tsubouchi, Amanda G. Fisher -- H3K9/HP1 and Polycomb; two key epigenetic silencing pathways for gene regulation and embryo development/Peter Nestorov, Mathieu Tardat, Antoine H.F.M. Peters -- Parental epigenetic asymmetry in mammals/Rachel Duffié, Déborah Bourc'his.
Subjects	Epigenetics.
Notes	Includes bibliographical references and index.
Series	Current topics in developmental biology, 0070-2153; v. 104
	Current topics in developmental biology; v. 104.

Epigenetics and human health; linking hereditary, environmental, and nutritional aspects

LCCN	2012405491
Type of material	Book
Main title	Epigenetics and human health; linking hereditary, environmental, and nutritional aspects/edited by Alexander G. Haslberger, co-edited by Sabine Gressler.
Published/Created	Weinheim; Wiley-VCH, c2010.
Description	xxi, 298 p.; ill.; 25 cm.
ISBN	9783527324279 (hbk.; alk. paper)
	3527324275 (hbk.; alk. paper)
LC classification	QH450 .E656 2010
Related names	Haslberger, Alexander G.
	Gressler, Sabine.

Bibliography

Contents

The research program in epigenetics: the birth of a new paradigm/Paolo Vineis -- Interactions between nutrition and health/Ibrahim Elmadfa -- Epigenetics: comments from an ecologist/Fritz Schiemer -- Interaction of hereditary and epigenetic mechanisms in the regulation of gene expression/Thaler Roman ... [et al.] -- Methylenetetrahydrofolate reductase C677T and A1298C polymorphisms and cancer risk: a risk review of the published meta-analyses/Stefania Boccia -- The role of biobanks for the understanding of gene-environment interactions/Christian Viertler, Michaela Mayrhofer, and Kurt Zatloukal -- Case studies on epigenetic inheritance/Gunnar Kaati -- Gentoxic, non-genotoxic and epigenetic mechanisms in chemical hepatocarcinigenesis: implications for safety evaluation/Wilfried Bursch -- Carcinogens in foods: occurrence, modes of action and modulation of human risks by genetic factors and dietary constituents/M. Mišík ... [et al.] -- From molecular nutrition to nutritional systems biology/Guy Vergères -- Effects of dietary natural compounds on DNA methylation related to cancer chemoprevention and anticancer epigenetic therapy/Barbara Maria Stefanska and Krystyna Fabianowska-Majewska -- Health determinants throughout the life cycle/Petra Rust -- Viral infections and epigenetic control mechanisms/Klaus R. Huber -- Epigenetics aspects in gyneacology and reproductive medicine/Alexander Just and Johannes Huber -- Epigenetics and tumorigenesis/Heidrun Karlic and Franz Varga -- Epigenetic approaches in oncology/Sabine Zöchbauer-Müller and Robert M. Mader -- Epigenetic dysregulation in aging and cancer/Despina Komninou and John P. Richie -- The impact of genetic and environmental factors in neurodegeneration: emerging role of epigenetics/Lucia Migliore and Fabio Coppedè -- Epigenetic biomarkers in neurodegenerative disorders/Borut Peterlin -- Epigenetic mechanisms in asthma/Rachel L. Miller and Julie Herbstman -- Public health genomics-integrating genomics and epigenetics into national and European health

	strategies and polices/Tobias Schulte in den Bäumen and Angela Brand -- Taking a first step: epigenetic health and responsibility/Astrid H. Gesche.
Subjects	Epigenesis.
	Health.
	Medical genetics.
	Genetic disorders.
	Environmental health.
	Epigenesis, Genetic.
	Environmental Health.
	Genetic Predisposition to Disease.
	Metagenomics.
	Nutrigenomics.
Notes	Includes bibliographical references and index.

Epigenetics; development and disease

LCCN	2012946199
Type of material	Book
Main title	Epigenetics; development and disease/Tapas K. Kundu, editor.
Published/Created	Dordrecht; New York; Springer, c2013.
Description	xxvi, 689 p.; ill. (some col.); 24 cm.
ISBN	9789400745247 (alk. paper)
	9400745249 (alk. paper)
	9789400745254 (ebk.)
LC classification	QH450 .E657 2013
	QH611 .S84 v.61 2013
Related names	Kundu, Tapas K.
Contents	Chromatin organization, epigenetics and differentiation: an evolutionary perspective -- Secondary structures of the core histone N-terminal tails: their role in regulating chromatin structure -- Megabase replication domains along the human genome: relation to chromatin structure and genome organisation -- Role of DNA methyltransferases in epigenetic regulation in bacteria -- Metabolic aspects of epigenome: coupling of S-adenosylmethionine synthesis and gene regulation on chromatin by SAMIT module -- Epigenetic regulation of male germ cell differentiation -- Epigenetic regulation of skeletal muscle development and differentiation -- Small

changes, big effects: chromatin goes aging -- Homeotic gene regulation: a paradigm for epigenetic mechanisms underlying organismal development -- Basic mechanisms in RNA polymerase I transcription of the ribosomal RNA genes -- The RNA polymerase II transcriptional machinery and its epigenetic context -- RNA polymerase III transcription-regulated by chromatin structure and regulator of nuclear chromatin organization -- The role of DNA methylation and histone modifications in transcriptional regulation in humans -- Histone variants and transcription regulation -- Noncoding RNAs in chromatin organization and transcription regulation: an epigenetic view -- Chromatin structure and organization: the relation with gene expression druing development and disease -- Cancer: an epigenetic landscape -- Epigenetic regulation of cancer stem cell gene expression -- Role of epigenetic mechanisms in the vascular complications of diabetes -- Epigenetic changes in infammatory and autoimmune diseases -- Epigenetic regulation of HIV-1 persistence and evolving strategies for virus eradication -- Epigenetics in Parkinson's and Alzheimer's diseases -- Cellular redox, epigenetics and diseases -- Stem cell plasticity in development and cancer: epigenetic origin of cancer stem cells -- Histone acetylation as a therapeutic target -- DNA methylation and cancer -- Role of epigenetics in inflammation-associated diseases -- Plasmodium falciparum: epigenetic control of var gene regulation and disease.

Subjects Epigenetics.
Chromatin--Genetics.
Epigenesis, Genetic.
Epigenomics.
Chromatin--genetics.
Gene Expression Regulation.
Notes Includes bibliographical references and index.
Series Subcellular biochemistry, 0306-0225; v. 61
Sub-cellular biochemistry; v.61. 0306-0225

Epigenetics, environment, and genes

LCCN	2012951950
Type of material	Book
Main title	Epigenetics, environment, and genes/edited by Sun Woo Kang, MD, PhD.
Published/Produced	Toronto; New Jersey; Apple Academic Press, [2013] ©2013
Description	xxvi, 306 pages; illustrations; 24 cm
ISBN	9781926895253 (hardcover; acid-free paper)
	1926895258 (hardcover; acid-free paper)
LC classification	QH450 .E6574 2013
Related names	Kang, Sun Woo, editor.
Contents	Pt. I. From genes to epigenomes. The epigenetic perspective/Albert H.C. Wong, Irving I. Gottesman, and Arturas Petronis -- Correlating CpG islands, motifs, and sequence variants/Leah Spontaneo and Nick Cercone -- pt. II. DNA methylation. Promoter hypermethylation/Davide Pellacani, Richard J. Packer, Fiona M. Frame, Emma E. Oldridge, Paul A. Berry, Marie-Christine Labarthe, Michael J. Stower, Matthew S. Simms, Anne T. Collins, and Norman J. Maitland -- Genome-wide mapping/Ryan K.C. Yuen, Ruby Jiang, Maria S. Peñaherrera, Deborah E. McFadden, and Wendy P. Robinson -- pt. III. Histone modifications. Gene silencing/Irina A. Maksakova, Preeti Goyal, Jörn Bullwinkel, Jeremy P. Brown, Misha Bilenky, Dixie L. Mager, Prim B. Singh, and Matthew C. Lorincz -- Histone (H1) phosphorylation/Anna Gréen, Bettina Sarg, Henrik Gréen, Anita Lönn, Herbert H. Lindner, and Ingemar Rundquist -- Post-translational modifications/Stephanie D. Byrum, Sean D. Taverna, and Alan J. Tackett -- pt. IV. Chromatin modifications. Chromatin signature/Brandon J. Thomas, Eric D. Rubio, Niklas Krumm, Pilib Ó. Broin, Karol Bomsztyk, Piri Welcsh, John M. Greally, Aaron A. Golden, and Anton Krumm -- Bivalent chromatin modification/Marco De Gobbi, David Garrick, Magnus Lynch, Douglas Vernimmen, Jim R. Hughes, Nicolas Goardon, Sidinh Luc, Karen M. Lower, Jacqueline A. Sloane-Stanley, Cristina Pina,

	Shamit Soneji, Raffaele Renella, Tariq Enver, Stephen Taylor, Sten Eirik W. Jacobsen, Paresh Vyas, Richard J. Gibbons, and Douglas R. Higgs -- pt. V. The role of environment during evolution. Adaptive divergence/Nora Khaldi and Denis C. Shields.
Subjects	Epigenetics.
	Gene expression.
	Genes.
	Chromatin.
	Nature and nurture.
	Epigenomics.
	Chromatin.
	Epigenesis, Genetic.
	Evolution, Molecular.
	Épigénétique.
	Expression génique.
	Gènes.
	Chromatine.
	Hérédité et milieu.
	Epigenomics.
	Chromatin.
	Epigenesis, Genetic.
	Evolution, Molecular.
	Chromatin.
	Epigenetics.
	Gene expression.
	Genes.
	Nature and nurture.
	Epigenetik.
	Épigénétique.
	Expression génique.
	Chromatine.
	Hérédité et milieu.
Notes	"CRC Press, Taylor & Francis Group"--Cover.
	Includes bibliographical references and index.
Additional formats	Kang, Sun Woo. Epigenetics, Environment, and Genes. Hoboken; CRC Press, 2013.

Epigenetics; mechanisms, functions and human effects
LCCN   2009032909

Type of material	Book
Main title	Epigenetics; mechanisms, functions and human effects/Balázs Pintér and Zsolt Mészáros, editors.
Published/Created	New York; Nova Science Publishers, c2010.
Description	xiii, 303 p.; ill. (some col.); 27 cm.
ISBN	9781607414544 (hardcover)
	1607414546 (hardcover)
LC classification	RC268.48 .E65 2010
Related names	Pintér, Balázs.
	Mészáros, Zsolt.
Summary	"Epigenetics refers to DNA and chromatin modifications that persist from one cell division to the next despite a lack of change in the underlying DNA sequence. It is an emergent field since its implication in physiological and disease processes is now widely accepted. The effects of epigenetics play an important role in development but can also arise stochastically with aging. Since their discovery in 1983, cancer-associated epigenetic changes have become accepted as playing a pivotal role in carcinogenesis, in conjunction with classical genomic mutations. Thus it is now well established that an epigenetic mechanism underlies the pathogenesis of cancer and there is evidence implicating epigenetic factors in the pathogenesis of idiopathic mental disorders such as schizophrenia, bipolar disorder and even autism. This new book gathers the latest research from around the globe in this field of study and related topics."--Publisher's description.
Contents	Techniques in cancer epigenetics/Michele Cummings and Laura Smith -- Melatonin; epigenetic mechanisms in cancer inhibition/Ahmet Korkmaz and Russel J. Reiter -- The epigenetic (re)programming of phenotypic differences in behavior/Patrick O. McGowan, Michael J. Meaney and Moshe Szyf -- Role of H3 phosphorylation in different biological processes in normal and disease states/Geneviève P. Delcuve, Paula S. Espino and James R. Davie -- The role of DNA methylation in colorectal cancer/Ivana Fridrichova -- Epigenetic clues to arthritic diseases/Helmtrud I. Roach and Emma L. Williams -- Evolution of the Fpg/Nei family of DNA glycosylases/Dorothy E. Pumo ... [et al.] -- The clinical

	and etiological relevance of MLH1 promoter hypermethylation in ovarian tumors/Tuya Pal and O. Thomas Mueller -- Epigenetic role in longevity echoes conserved stress survival strategies/Joan Smith Sonneborn -- Epigenetics and neurodegeneration; a connection overlooked/Sueli C.F. Marques, Clâaudia M.F. Pereira and Tiago F. Outeiro -- DNA methylation changed by 5-azacytidine can affect normal fetal and placental development/Ljiljana Serman ... [et al.] -- Epigenetic analysis in breast cancer progression/Frank A. Orlando, Lisa M. Dyer and Kevin D. Brown -- Cancer epigenetics/Frank A. Orlando, Berna Demircan and Kevin D. Brown.
Subjects	Cancer--Etiology.
	Cancer--Genetic aspects.
	Epigenesis.
	Neoplasms--etiology.
	Neoplasms--genetics.
	Epigenesis, Genetic.
Notes	Includes bibliographical references and index.
Series	Genetics--research and issues series

Epigenetics of aging

LCCN	2009932905
Type of material	Book
Main title	Epigenetics of aging/Trygve O. Tollefsbol, editor.
Published/Created	New York; Springer, c2010.
Description	xiv, 469 p. ill. (some col.); 24 cm.
ISBN	9781441906380 (acid-free paper)
	144190638X (acid-free paper)
	9781441906397 (e-ISBN)
	1441906398 (ebook)
LC classification	QP86 .E44 2010
Related names	Tollefsbol, Trygve O.
Subjects	Aging--Genetic aspects.
	Epigenesis.
	Genetic regulation.
	Aging--genetics.
	DNA Methylation--physiology.
	Epigenesis, Genetic.

Notes   Includes bibliographical references and index.

Evo-devo of child growth; treatise on child growth and human evolution

LCCN	2011037214
Type of material	Book
Personal name	Hochberg, Z.
Main title	Evo-devo of child growth; treatise on child growth and human evolution/Ze'ev Hochberg.
Published/Created	Hoboken, N.J.; Wiley-Blackwell, c2012.
Description	xii, 235 p.; ill.; 27 cm.
ISBN	9781118027165 (hardback) 1118027167 (hardback)
LC classification	RJ131 .H595 2012
Portion of title	Treatise on child growth and human evolution
Summary	"Bringing together the evolutionary theory of life history and principles from evo-devo, this book analyzes child growth and development in a presentation accessible to evolutionary biologists and clinicians. It shows how transitions between unique life-history phases relate to adaptive responses such as body composition, stature, and differences between male and female development. The book also explores how these transitions can illuminate the sequence and mechanisms of child growth, and serve as diagnostic tools for growth disorders. Researchers and students will find a broad range of theoretical and practical concepts discussed"-- Provided by publisher.
Contents	I. Introduction. A. Evolutionary thinking in medicine; B. Evo-devo; C. Life history theory; D. Evolutionary perspective in child growth and maturation; E. Child growth and the environment; F. Heterochrony and allometry; G. Adaptive plasticity in life history -- II. Child growth and the theory of life history. A. Life-history stages; B. Transitions between life-history stages; C. Developmental plasticity and adaptation; D. Cultural adaptation to the environment; E. Adaptive plasticity of attachment behaviors; F. Note by George Chrousos on stress in early life; a developmental and evolutionary perspective. 1. Stress concepts; 2. Stress mechanisms; 3. Pathological effects of stress; G. Note by Stefan Borenstein and Andreas Androutsellis-Theotokis on

endogenous stem cells as components of plasticity and adaptation. 1. The adult mammalian brain; plastic or rigid?; 2. Hidden plasticity potential in the brain; 3. Neurogenic cell vs. neural stem cell; 4. Does the role of neural stem cells change from the developing age to the adult?; 5. The disconnect between neurogenesis and the presence of neural stem cells; 6. Fetal vs. adult neural stem cells; 7. Signal transduction of stem cell regulation; 8. Beyond the nervous system; 9. Conclusions -- III. Fetal growth. A. Endocrine and metabolic control of fetal growth; B. The role of the placenta; C. Developmental origins of health and adult disease (DOHaD); D. Imprinted genes and intrauterine growth; E. Note by Alan Templeton on the evolutionary connection between senescence and childhood growth and development. 1. An evolutionary theory of aging; 2. Thrifty genotypes and antagonistic pleiotropy; 3. Thrifty genotypes and heart disease; 4. Why we grow old; the answer -- IV. Infancy. A. The reproductive dilemma; B. The obstetrical dilemma; C. Growth of the infant; D. Endocrine aspects of infantile growth; E. Infancy -- childhood transition; determination of adult stature; F. Weaning from breast-feeding -- V. Childhood. A. The weanling's dilemma; B. The grandmother theory; C. Growth of the child; D. Endocrine aspects of childhood growth -- VI. Juvenility. A. The social/cognitive definition of juvenility; B. Paleoanthropological juvenility and teeth eruption; C. Adrenarche; D. Juvenile body composition; E. Growth of the juvenile; F. Trade-offs for the timing of transition to juvenility; G. Precocious juvenility; H. The Pygmy paradigm for precocious juvenility; I. Evolutionary perspective in precocious juvenility -- VII. Adolescence. A. Human evolution of adolescence; B. Transition from juvenility to adolescence; C. Pubertal growth -- VIII. Youth -- IX. Evolutionary strategies for body size. A. The little people of Flores; B. Lessons from the great apes; C. The handicap theory; D. Sexual dimorphism; E. The role of sex steroids -- X. Energy considerations. A. Endocrine control of energy expenditure; B. Weaning and growth in a malnourished environment -- XI. Stage transitions; trade-offs and adaptive phenotypic plasticity.

	A. Transgenerational influences in life-stage transition; B. Epigenetics and life-history stage transitions; C. Note by Ken Ong on population genetics and child growth and maturation. 1. Genetic adaptation; 2. The genetic epidemiology of child growth and maturation; 3. Basic principles and heritability estimates from twin studies; 4. More complex heritability models; 5. Heritability is dependent upon the setting; 6. Essential genes for childhood growth and maturation; 7. Common genetic variants for childhood growth and maturation; 8. GWAS findings lead to new biology; 9. GWAS findings lead to new phenotypic understanding; 10. Genetic adaptations for childhood growth and maturation; 11. Conclusions; D. Note by Moshe Szyf on the DNA methylation pattern as a molecular link between early childhood and adult health. 1. Introduction; 2. DNA methylation patterns and their roles in cellular differentiation and gene expression; 3. DNA methylation as a genome adaptation mechanism; 4. Epigenetic programming by the early life social environment; 5. Genome and system-wide impact of early life adversity; 6. Prospective and summary -- XII. Life history theory in understanding growth disorders. A. Down syndrome; B. Noonan's syndrome; C. Silver-Russell syndrome; D. Additional cases -- XIII. When the packages disintegrate -- XIV. Concluding remarks.
Subjects	Child development. Children--Growth. Human evolution. Child Development. Biological Evolution. Growth.
Notes	Includes bibliographical references and index.

Formalin-fixed paraffin-embedded tissues; methods and protocols

LCCN	2011922256
Type of material	Book
Main title	Formalin-fixed paraffin-embedded tissues; methods and protocols/edited by Fahd Al-Mulla.
Published/Created	New York; Humana Press, c2011.
Description	xii, 312 p.; ill.; 27 cm.
ISBN	9781617790546 (alk. paper)

	1617790540 (alk. paper)
	9781617790553 (e-ISBN)
	1617790559 (e-ISBN)
LC classification	QP88 .F68 2011
	QH506 .M45 v.724
Related names	Al-Mulla, Fahd.
Contents	Regulatory and ethical Issues on the utilization of FFPE tissues in research/Catherine M. With, David L. Evers, and Jeffrey T. Mason -- Tissue microarrays; construction and uses/Carol B. Fowler ... [et al.] -- Standardization in immunohistology/Anthony S.-Y. Leong and Trishe Y.-M. Leong -- Multiple immunofluorescence labeling of formalin-fixed paraffin-embedded tissue/David Robertson and Clare M. Isacke -- Microwaves for chromogenic In situ hybridization/Anthony S.-Y. Leong and Zenobia Haffajee -- Automated analysis of FISH-stained HER2/neu samples with metafer/Christian Schunck and Eiman Mohammad -- Laser capture microdissection of FFPE tissue sections bridging the gap between microscopy and molecular analysis/Renate Burgemeister -- Nucleic acids extraction from laser microdissected FFPE tissue sections/Renate Burgemeister -- Microarray-based CGH and copy number analysis of FFPE samples/Fahd Al-Mulla -- Microarray profiling of DNA extracted from FFPE tissues using SNP 6.0 affymetrix platform/Marianne Tuefferd ... [et al.] -- Whole genome amplification of DNA extracted from FFPE tissues/Mira Bosso and Fahd Al-Mulla -- Pyrosequencing of DNA extracted from formalin-fixed paraffin-embedded tissue/Brendan Doyle, Ciaran O'Riain, and Kim Appleton -- Analysis of DNA methylation in FFPE tissues using the methylight technology/Ashraf Dallol ... [et al.] -- RT-PCR gene expression profiling of RNA from paraffin-embedded tissues prepared using a range of different fixatives and conditions/Mei-Lan Liu ... [et al.] -- RT-PCR-based gene expression profiling for cancer biomarker discovery from fixed, paraffin-embedded tissues/Aaron Scott ... [et al.] -- MicroRNA isolation from formalin-fixed, paraffin-embedded tissues/Aihua Liu and Xiaowei Xu -- Gene expression profiling of RNA extracted from FFPE

146 Bibliography

	tissues; NuGEN technologies' whole-transcriptome amplification system/Leah Turner, Joe Don Heath, and Nurith Kurn -- Protein mass spectrometry applications on FFPE tissue sections/Carol B. Fowler, Timothy J. O'Leary, and Jeffrey T. Mason -- Alternative fixative to formalin fixation for molecular applications; the RCL2®-CS100 approach/Amelie Denouel ... [et al.].
Subjects	Tissues--Analysis--Laboratory manuals.
	Tissue culture--Laboratory manuals.
	Diagnostic immunohistochemistry--Laboratory manuals.
	Paraffin Embedding--methods--Laboratory Manuals.
	Tissue Culture Techniques--Laboratory Manuals.
	Gene Expression Profiling--Laboratory Manuals.
	Tissue Array Analysis--methods--Laboratory Manuals.
	Tissue culture techniques.
	Tissue culture techniques--Ethics.
	Tissue array analysis--Methods.
	Paraffin embedding--Methods.
Notes	Includes bibliographical references and index.
Series	Springer protocols
	Methods in molecular biology, 1064-3745; 724
	Springer protocols.
	Methods in molecular biology (Clifton, N.J.); v. 724.
	1064-3745

Functional analysis of DNA and chromatin
LCCN	2013950902
Type of material	Book
Main title	Functional analysis of DNA and chromatin/edited by Juan C. Stockert, Jesús Espada, Alfonso Blázquez-Castro.
Published/Created	New York; Humana Press; Springer, c2014.
Description	xv, 365 p.; ill. (some col.); 26 cm.
ISBN	9781627037051 (alk. paper)
	1627037055 (alk. paper)
LC classification	QP624 .F86 2014
Related names	Stockert, Juan C.
	Espada, Jesús.
	Blázquez-Castro, Alfonso.
Contents	Predictive binding geometry of ligands to DNA minor groove; isohelicity and hydrogen-bonding

pattern/Juan C. Stockert -- Using microchip gel electrophoresis to probe DNA-drug binding interactions/Nan Shi and Victor M. Ugaz -- Identifying different types of chromatin using Giemsa staining/Juan C. Stockert, Alfonso Blázquez-Castro, and Richard W. Horobin -- Analysis of DNA damage and repair by comet fluorescence in situ hybridization (comet-FISH)/Michael Glei and Wiebke Schlörmann -- Alkaline nuclear dispersion assays for the determination of DNA damage at the single cell level/Piero Sestili and Carmela Fimognari -- Polarization microscopy of extended chromatin fibers/Maria Luiza S. Mello and Benedicto de Campos Vidal -- DNA labeling in vivo; quantification of epidermal stem cell chromatin content in whole mouse hair follicles using Fiji image processing software/Elisa Carrasco, María I. Calvo, and Jesús Espada -- Historical overview of bromo-substituted DNA and sister chromatid differentiation/Roberto Mezzanotte and Mariella Nieddu -- Image analysis of chromatin remodelling/Benedicto de Campos Vidal, Marina B. Felisbino, and Maria Luiza S. Mello -- FISH methods in cytogenetic studies/Miguel Pita ... [et al.] -- Ultrastructural and immunofluorescent methods for the study of the XY body as a biomarker/Roberta B. Sciurano and Alberto J. Solari -- Atomic force microscopy for analyzing metaphase chromosomes; comparison of AFM images with fluorescence labeling images of banding patterns/Osamu Hoshi and Tatsuo Ushiki -- Selective detection of phagocytic phase of apoptosis in fixed tissue sections/Vladimir V. Didenko -- Selective transport of cationized fluorescent topoisomerase into nuclei of live cells for DNA damage studies/Candace L. Minchew and Vladimir V. Didenko -- Visualization and interpretation of eukaryotic DNA replication intermediates in vivo by electron microscopy/Kai J. Neelsen ... [et al.] -- Combined bidimensional electrophoresis and electron microscopy to study specific plasmid DNA replication intermediates in human cells/Cindy Follonier and Massimo Lopes --

Standard DNA methylation analysis in mouse epidermis; bisulfite sequencing, methylation-specific PCR, and 5-methyl-cytosine (5mC) immunological detection/Jesús Espada, Elisa Carrasco, and María I. Calvo -- Methyl-combing; single-molecule analysis of DNA methylation on stretched DNA fibers/Attila Németh -- Investigating 5-hydroxymethylcytosine (5hmC); the state of the art/Colm E. Nestor ... [et al.] -- Hydroxymethylated DNA immunoprecipitation (hmeDIP)/Colm E. Nestor and Richard R. Meehan -- Microscale thermophoresis for the assessment of nuclear protein-binding affinities/Wei Zhang ... [et al.] -- Analysis of histone posttranslational modifications from nucleolus-associated chromatin by mass spectrometry/Stefan Dillinger ... [et al.] -- Salt-urea, sulfopropyl-sepharose, and covalent chromatography methods for histone isolation and fractionation/Pedro Rodriguez-Collazo ...[et al.] -- Chromatin immunoprecipitation/Javier Rodríguez-Ubreva and Esteban Ballestar -- Analysis of chromatin composition of repetitive sequences; the ChIP-Chop assay/Raffaella Santoro -- Purification of specific chromatin domains from single- copy gene loci in Saccharomyces cerevisiae/Stephan Hamperl ... [et al.] -- Deep sequencing of small chromatin-associated RNA; isolation and library preparation/Sarah Daniela Diermeier, Thomas Schubert, and Gernot Längst -- Deep sequencing of small chromatin-associated RNA; bioinformatic analysis/Sarah Daniela Diermeier and Gernot Längst.

Subjects   DNA--Analysis--Laboratory manuals.
Chromatin--Analysis--Laboratory manuals.
Chromatin--Laboratory Manuals.
DNA--analysis--Laboratory Manuals.

Notes   Includes bibliographical references and index.

Series   Methods in molecular biology, 1064-3745; 1094
Springer protocols
Methods in molecular biology (Clifton, N.J.); v. 1094.
1064-3745
Springer protocols (Series) 1949-2448

Fusarium; genomics, molecular and cellular biology

LCCN	2013444007
Type of material	Book
Main title	Fusarium; genomics, molecular and cellular biology/edited by Daren W. Brown and Robert H. Proctor
Published/Created	Norfolk, UK; Caister Academic Press, c2013.
Description	182, 2 p. of plates; ill. (some col.); 26 cm.
ISBN	9781908230256
	1908230258
LC classification	QK625.T8 F86 2013
Related names	Brown, Daren W.
	Proctor, Robert H.
Contents	An overview of Fusarium/John F. Leslie and Brett A. Summerell -- Sex and fruiting in Fusarium/Frances Trail -- Structural dynamics of Fusarium genomes/H. Corby Kistler, Martijn Rep and Li-Jun Ma -- Molecular genetics and genomic approaches to explore Fusarium infection in wheat floral tissue/Martin Urban and Kim E. Hammond-Kosack -- Applying proteomics to investigate the interactions between pathogenic Fusarium species and their hosts/Linda J. Harris, Thérèse Ouellet and Rajagopal Subramaniam -- Repeat-induced point mutation, DNA methylation and heterochromatin in Gibberella zeae (Anamorph: Fusarium graminearum)/Kyle R. Pomraning, [et al.] -- The nitrogen regulation network and its impact on secondary metabolism and pathogenicity/Philipp Wiemann and Bettina Tudzynski -- Diversity of polyketide synthases in Fusarium/Daren W. Brown and Robert H. Proctor -- Plant responses to Fusarium metabolites/Takumi Nishiuchi.
Subjects	Fusarium.
Notes	Includes bibliographical references and index.

Gene discovery for disease models

LCCN	2010028355
Type of material	Book
Main title	Gene discovery for disease models/edited by Weikuan Gu and Yongjun Wang.

Published/Created	Hoboken, N.J.; Wiley, c2011.
Description	xiii, 537 p., [16] p. of plates; ill. (some col.); 25 cm.
ISBN	9780470499467 (cloth)
LC classification	RB155 .G3584 2011
Related names	Gu, Weikuan.
Contents	Introduction; gene discovery-from positional cloning to genomic cloning -- High throughput gene expression analysis and the identification of expression QTLs -- DNA methylation in the pathogenesis of autoimmunity -- Ccell-based analysis with microfluidic chip -- Missing dimension; protein turnover rate measurement in gene discovery -- Bioinformatics tools for the prediction of gene function -- Determination of genomic locations of targeted genetic loci -- Mutation discovery using high throughput mutation screening technology -- Candidate screening through gene expression profile -- Candidate screening through high-density SNP array -- Gene discovery through direct genome sequencing -- Candidate screening through bioinformatics tools -- Using an integrative strategy to identify mutations -- Determination of the function of a mutant in a gene -- Confirmation of a mutation by multiple molecular approaches -- Confirmation of a mutation by microRNA -- Confirmation of function of a gene by translational approaches -- Confirmation of single nucleotide mutations -- Initial identification and confirmation of a QTL gene -- Gene discovery of crop diseases in the post genome era -- Impact of whole genome genetic element analysis on gene discovery of disease models -- Impact of whole genome protein analysis on gene discovery of disease models.
Subjects	Medical genetics.
	Mutation (Biology)
	Genomics.
	Genetic disorders.
	Genetic Association Studies--methods.
	Models, Genetic.
	Mutation.
Notes	Includes bibliographical references and index.

Genes and the motivation to use substances

LCCN	2014940945
Type of material	Book
Main title	Genes and the motivation to use substances/Scott F. Stoltenberg, editor.
Published/Produced	New York; Heidelberg; Springer Verlag, [2014] ©2014
Description	ix, 144 pages; illustrations (some color); 24 cm.
Links	Cover http://swbplus.bsz-bw.de/bsz403314046cov.htm 20140331142738
ISBN	9781493906529 (alk. paper)
	1493906526 (alk. paper)
LC classification	RC564 .G3729 2014
Related names	Stoltenberg, Scott F., editor.
	Nebraska Symposium on Motivation (61st; 2014; Lincoln, Neb.)
Contents	Do genes motivate substance use?/Scott F. Stoltenberg -- Rodent models of genetic contributions to motivation to abuse alcohol/John C. Crabbe -- The adolescent origins of substance use disorders; a behavioral genetic perspective/Matt McGue, Dan Irons and William G. Iacono -- Genes, brain, behavior and context; the developmental matrix of addictive behavior/Robert A. Zucker -- Have the genetics of cannabis involvement gone to pot?/Arpana Agrawal and Michael T. Lynskey -- The DNA methylation signature of smoking; an archetype for the identification of biomarkers for behavioral illness/Robert A. Philibert, S.R.H. Beach and Gene H. Brody -- Genomics of impulsivity; integrating genes and neuroscience/David Goldman.
Subjects	Substance abuse--Psychological aspects.
	Genetic psychology.
	Psychology.
	Biological Psychology.
	Cognitive Psychology.
	Human Genetics.
	Genetic psychology.
	Psychology.
	Substance abuse--Psychological aspects.
	Angewandte Psychologie
	Humangenetik

Subject keywords	Umwelt Abhängigkeit Rauschgift Motivation
Notes	Includes bibliographical references and index.
Series	Nebraska Symposium on Motivation, 0146-7875; volume 61 Nebraska Symposium on Motivation. Nebraska Symposium on Motivation; v. 61.

Genomic imprinting; methods and protocols

LCCN	2012943828
Type of material	Book
Main title	Genomic imprinting; methods and protocols/edited by Nora Engel.
Published/Created	New York; Humana Press/Springer, c2012.
Description	xi, 297 p.; ill. (some col.); 26 cm.
ISBN	9781627030106 (alk. paper) 1627030107 (alk. paper) 9781627030113 (eBook) 1627030115 (eBook)
LC classification	QH450 .G468 2012
Related names	Engel, Nora.
Contents	Uniparental embryos in the study of genomic imprinting/Yong Cheng, Dasari Amarnath, and Keith E. Latham -- Derivation of induced pluripotent stem cells by retroviral gene transduction in mammalian species/Masanori Imamura ... [et al.] -- Generation of trophoblast stem cells/Michael C. Golding -- Immunomagnetic purification of murine primordial germ cells/Emily Y. Smith and James L. Resnick -- Whole genome methylation profiling by immunoprecipitation of methylated DNA/Andrew J. Sharp -- Identification of imprinted loci by transcriptome sequencing/Tomas Babak -- Data mining as a discovery tool for imprinted genes/Chelsea Brideau and Paul Soloway -- Engineering of large deletions and duplications in vivo/Louis Lefebvre -- Methylated DNA immunoprecipitation (MeDIP) from low amounts of cells/Julie Borgel, Sylvain Guibert, and Michael

	Weber -- Chromatin immunoprecipitation to characterize the epigenetic profiles of imprinted domains/Purnima Singh and Piroska E. Szabo -- Quantitative chromosome conformation capture/Raffaella Nativio, Yoko Ito, and Adele Murrell -- Genome-wide analysis of DNA methylation in low cell numbers by reduced representation bisulfite sequencing/Sebastien A. Smallwood and Gavin Kelsey -- Isolation of RNA and DNA from single preimplantation embryos and a small number of mammalian oocytes for Imprinting Studies/Sarah Rose Huffman, Md Almamun, and Rocio Melissa Rivera -- Generation of cDNA libraries from RNP-derived regulatory noncoding RNAs/Mathieu Rederstorff -- Co-immunoprecipitation of long noncoding RNAs/Victoria A. Moran, Courtney N. Niland, and Ahmad M. Khalil -- Specialized technologies for epigenetics in plants/Wenyan Xiao -- Computational studies of imprinted genes/Martina Paulsen -- Insights on imprinting from beyond mice and men/Andrew Pask -- Nonmammalian parent-of-origin effects/Elena de la Casa-Esperon.
Subjects	Genomic imprinting--Laboratory manuals.
	Genomic Imprinting--Laboratory Manuals.
Form/Genre	Laboratory Manuals.
Notes	Includes bibliographical references and index.
Series	Methods in molecular biology, 1064-3745; 925
	Springer protocols
	Methods in molecular biology (Clifton, N.J.); v. 925. 1064-3745
	Springer protocols. 1949-2448

Germline development; methods and protocols
LCCN	2011941357
Type of material	Book
Main title	Germline development; methods and protocols/edited by Wai-Yee Chan, Le Ann Blomberg.
Published/Created	New York; Humana Press, c2012.
Description	xi, 256 p.; ill. (some col.); 26 cm.

Links	Publisher description http://www.loc.gov/catdir/enhancements/fy1316/2011941357-d.html Table of contents only http://www.loc.gov/catdir/enhancements/fy1316/2011941357-t.html
ISBN	9781617794353 (alk. paper) 161779435X (alk. paper) 9781617794360 (e-ISBN) 1617794368 (e-ISBN)
LC classification	QL964 .G47 2012
Related names	Chan, Wai-Yee. Blomberg, Le Ann.
Contents	Isolation of fetal gonads from embryos of timed-pregnant mice for morphological and molecular studies -- Neonatal testicular gonocytes isolation and processing for immunocytochemical analysis -- Isolation of undifferentiated and early differentiating type a spermatogonia from pou5f1-GFP reporter mice -- Isolation of human male germ-line stem cells using enzymatic digestion and magnetic-activated cell sorting -- Isolation and purification of murine male germ cells -- Preparation of enriched mouse syncitia-free pachytene spermatocyte cell suspensions -- Revealing the transcriptome landscape of mouse spermatogonial cells by tiling microarray -- Biochemical characterization of a testis-predominant isoform of n-[alpha] acetyltransferase -- Identification of novel long noncoding RNA transcripts in male germ cells -- Methylation profiling using methylated DNA immunoprecipitation and tiling array hybridization -- Spermatogenesis in cryptorchidism -- In vitro culture of fetal ovaries: A model to study factors regulating early follicular development -- Microspread ovarian cell preparations for the analysis of meiotic prophase progression in oocytes with improved recovery by cytospin centrifugation -- In vitro maturation (IVM) of porcine oocytes -- Experimental approaches to the study of human primordial germ cells -- Investigating the origins of somatic cell populations in the perinatal mouse ovaries using genetic lineage tracing and immunohistochemistry -- DNA methylation analysis of germ cells by using bisulfite-based sequencing methods -- Gene expression in mouse oocytes by RNA-Seq.

Subjects	Germ cells--Laboratory manuals.
	Germ Cells--growth & development--Laboratory Manuals.
	Cytological Techniques—methods--Laboratory Manuals.
Notes	Includes bibliographical references and index.
Series	Methods in molecular biology, 1064-3745; 825
	Springer protocols
	Methods in molecular biology (Clifton, N.J.); v. 825.
	1064-3745
	Springer protocols (Series)

High-throughput next generation sequencing; methods and applications

LCCN	2011923657
Type of material	Book
Main title	High-throughput next generation sequencing; methods and applications/edited by Young Min Kwon, Steven C. Ricke.
Published/Created	New York; Springer, c2011.
Description	xi, 308 p.; ill. (some col.) 27 cm.
Links	Publisher description
	http://www.loc.gov/catdir/enhancements/fy1111/201192 3657-d.html
	Table of contents only
	http://www.loc.gov/catdir/enhancements/fy1111/201192 3657-t.html
ISBN	9781617790881
	1617790885
	9781617790898 (e-isbn)
	1617790893 (e-isbn)
LC classification	QH447 .H54 2011
Related names	Kwon, Young Min.
	Ricke, Steven C., 1957-
Contents	Helicos single-molecule sequencing of bacterial genomes/Kathleen E. Steinmann ... [et al.] -- Whole-genome sequencing of unculturable bacterium using whole-genome amplification/Yuichi Hongoh and Atsushi Toyoda -- RNA sequencing and quantitation using the helicos genetic analysis system/Tal Raz ... [et al.] -- Transcriptome profiling using single-molecule direct RNA sequencing/Fatih Ozsolak and Patrice M. Milos -- Discovery of bacterial sRNAs by high-throughput

sequencing/Jane M. Liu and Andrew Camilli -- Identification of virus encoding microRNAs using 454 FLX sequencing platform/Byung-Whi Kong -- Ribosomal RNA depletion for massively parallel bacterial RNA-sequencing applications/Zhoutao Chen and Xiaoping Duan -- Integrating high-throughput pyrosequencing and quantitative real-time PCR to analyze complex microbial communities/Husen Zhang ... [et al.] -- Tag-encoded FLX amplicon pyrosequencing for the elucidation of microbial and functional gene diversity in any environment/Yan Sun, Randall D. Wolcott, and Scot E. Dowd -- Pyrosequencing of chaperonin-60 (cpn60) amplicons as a means of determining microbial community composition/John Schellenberg ... [et al.] -- Prescreening of microbial populations for the assessment of sequencing potential/Irene B. Hanning and Steven C. Ricke -- Metagenomics/Jack A. Gilbert ... [et al.] -- Metagenomic analysis of intestinal microbiomes in chickens/Taejoong Kim and Egbert Mundt -- Gene expression profiling; metatranscriptomics/Jack A. Gilbert and Margaret Hughes -- High-throughput insertion tracking by deep sequencing for the analysis of bacterial pathogens/Sandy M.S. Wong ... [et al.] -- Determining DNA methylation profiles using sequencing/Suhua Feng ... [et al.] -- Preparation of next-generation sequencing libraries using Nextera technology; simultaneous DNA fragmentation and adaptor tagging by in vitro transposition/Nicholas Caruccio -- Amplification-free library preparation for paired-end illumina sequencing/Iwanka Kozarewa and Daniel J. Turner -- Target-enrichment through amplification of hairpin-ligated universal targets for next-generation sequencing analysis/Pallavi Singh, Rajesh Nayak, and Young Min Kwon -- 96-plex molecular barcoding for the illumina genome analyzer/Iwanka Kozarewa and Daniel J. Turner.

Subjects Genomes--Analysis.
Gene expression--Analysis.
Gene mapping--Laboratory manuals.
Genome.
Gene Expression Profiling.
Metagenomics.

Notes	Includes bibliographical references and index.
Additional formats	Online version: High-throughput next generation sequencing. New York, N.Y.; Humana Press, c2011 (OCoLC)712789122
Series	Methods in molecular biology, 1064-3745; 733 Springer protocols Methods in molecular biology (Clifton, N.J.); v. 733 Springer protocols.

Imaging and tracking stem cells; methods and protocols

LCCN	2013941994
Type of material	Book
Main title	Imaging and tracking stem cells; methods and protocols/edited by Kursad Turksen, Ottawa Hospital Research Institute, Sprott Centre for Stem Cell Research, Regenerative Medicine Program, Smyth road 501, K1Y 8L6 Ottawa, Ontario, Canada.
Published/Produced	New York; Humana Press, [2013] ©2013
Description	xi, 218 pages; illustrations (mostly colored); 26 cm.
ISBN	9781627035583 (alk. paper) 1627035583 (alk. paper)
LC classification	QH588.S83 I43 2013
Related names	Turksen, Kursad, editor. Springer Science+Business Media, copyright holder.
Contents	Primary Culture and Live Imaging of Adult Neural Stem Cells and Their Progeny/Felipe Ortega, Benedikt Berninger, and Marcos R. Costa -- Labeling and Tracking of Human Mesenchymal Stem Cells Using Near-Infrared Technology/Marie-Therese Armentero, Patrizia Bossolasco, and Lidia Cova -- High Content Imaging and Analysis of Pluripotent Stem Cell-Derived Cardiomyocytes/Gábor Földes and Maxime Mioulane -- A High Content Imaging-Based Approach for Classifying Cellular Phenotypes/Joseph J. Kim, Sebastián L. Vega, and Prabhas V. Moghe -- Conversion of Primordial Germ Cells to Pluripotent Stem Cells: Methods for Cell Tracking and Culture Conditions/Go Nagamatsu and Toshio Suda -- Imaging and Tracking of Bone-Marrow-Derived Immune and Stem Cells/Youbo Zhao [and four

others] -- Co-Visualization of Methylcytosine, Global DNA, and Protein Biomarkers for In Situ 3-D DNA Methylation Phenotyping of Stem Cells/Jian Tajbakhsh -- Noninvasive Imaging of Myocardial Blood Flow Recovery in Response to Stem Cell Intervention/HuaLei Zhang and Rong Zhou -- Live Imaging of Early Mouse Embryos Using Fluorescently Labeled Transgenic Mice/Takaya Abe, Shinichi Aizawa, and Toshihiko Fujimori -- Live Imaging, Identifying, and Tracking Single Cells in Complex Populations In Vivo and Ex Vivo/Minjung Kang [and four others] -- Quantitative Evaluation of Stem Cell Grafting in the Central Nervous System of Mice by In Vivo Bioluminescence Imaging and Postmortem Multicolor Histological Analysis/Kristien Reekmans [and eight others] -- Micro-CT Technique for Three Dimensional Visualization of Human Stem Cells/Andrea Farini [and four others] -- Noninvasive Multimodal Imaging of Stem Cell Transplants in the Brain Using Bioluminescence Imaging and Magnetic Resonance Imaging/Annette Tennstaedt [and three others] -- Magnetic Resonance Imaging and Tracking of Stem Cells/Hossein Nejadnik, Rostislav Castillo, and Heike E. Daldrup-Link -- Whole Body MRI and Fluorescent Microscopy for Detection of Stem Cells Labeled with Superparamagnetic Iron Oxide (SPIO) Nanoparticles and DiI Following Intramuscular and Systemic Delivery/Boris Odintsov, Ju Lan Chun, and Suzanne E. Berry -- Molecular Imaging and Tracking Stem Cells in Neurosciences/Toma Spiriev, Nora Sandu, and Bernhard Schaller -- Bioluminescence Imaging of Human Embryonic Stem Cells-Derived Endothelial Cells for Treatment of Myocardial Infarction/Weijun Su [and four others].

Subjects Stem cells.
Molecular probes.
Radioisotope scanning.
Stem Cells--radionuclide imaging--Laboratory Manuals.
Cell Tracking--Laboratory Manuals.

	Imaging, Three-Dimensional--methods--Laboratory Manuals.
Form/Genre	Laboratory Manuals.
Notes	Includes bibliographical references and index.
Series	Methods in molecular biology, methods and protocols, 1064-3745; 1052 Springer protocols Methods in molecular biology (Clifton, N.J.); v. 1052 Springer protocols (Series)

In situ hybridization protocols

LCCN	2014944535
Type of material	Book
Main title	In situ hybridization protocols/edited by Boye Schnack Nielsen, Bioneer A/S, Molecular Histology, Hørholm, Denmark.
Edition	Fourth edition.
Published/Produced	New York; Humana Press, [2014] ©2014
Description	xi, 275 pages; illustrations; 26 cm.
ISBN	9781493914586 (alk. paper) 1493914588 (alk. paper)
LC classification	QH452.8 .I54 2014
Related names	Nielsen, Boye Schnack, editor.
Contents	Fixation/permeabilization procedure for mRNA in situ hybridization of zebrafish whole-mount oocytes, embryos, and larvae/Ricardo Fuentes and Juan Fernández -- MicroRNA in situ hybridization on whole-mount preimplantation embryos/Karen Goossens, Luc Peelman, and Ann Van Soom -- Whole-mount in situ hybridization (WISH) optimized for gene expression analysis in mouse embryos and embryoid bodies/Eleni Dakou [and five others] -- Whole-mount in situ hybridization using DIG-labeled probes in planarian/Agnieszka Rybak-Wolf and Jordi Solana -- In situ hybridization on whole-mount zebrafish embryos and young larvae/Bernard Thisse and Christine Thisse -- LNA-based in situ hybridization detection of mRNAs in embryos/Diana K. Darnell and Parker B. Antin -- Chromogen detection of microRNA in frozen clinical tissue

samples using LNA probe technology/Boye Schnack Nielsen, Trine Møller, and Kim Holmstrøm -- MicroRNA in situ hybridization in tissue microarrays/Julia J. Turnock-Jones and John P.C. Le Quesne -- Fluorescence in situ hybridization for detection of small RNAs on frozen tissue sections/Asli Silahtaroglu -- Sensitive and specific in situ hybridization for early drug discovery/Pernille A. Usher ... [and six others] -- Zinc-based fixation for high-sensitivity in situ hybridization; a nonradioactive colorimetric method for the detection of rare transcripts on tissue sections/Electra Stylianopoulou, George Skavdis, and Maria Grigoriou -- Dual-color ultrasensitive bright-field RNA in situ hybridization with RNAscope/Hongwei Wang ... [and eight others] -- Fully automated fluorescence-based four-color multiplex assay for co-detection of microRNA and protein biomarkers in clinical tissue specimens/Lorenzo F. Sempere -- Multiplexed miRNA fluorescence in situ hybridization for formalin-fixed paraffin-embedded tissues/Neil Renwick ... [and three others] -- Simultaneous detection of nuclear and cytoplasmic RNA variants utilizing stellaris RNA fluorescence in situ hybridization in adherent cells/Sally R. Coassin ... [and three others] -- Quantitative ultrasensitive bright-field RNA in situ hybridization with RNAscope/Hongwei Wang ... [and eight others] -- Identification of low-expressing transcripts of the NPY receptors' family in the murine lingual epithelia/Sergei Zolotukhin -- In situ hybridization for fungal ribosomal RNA sequences in paraffin-embedded tissues using biotin-labeled locked nucleic acid probes/Kathleen T. Montone -- In situ hybridization freeze-assisted punches (IFAP); technique for liquid-based tissue extraction from thin slide-mounted sections for DNA methylation analysis/Lawrence S. Own and Paresh D. Patel -- miRNA detection at single-cell resolution using microfluidic LNA Flow-FISH/Meiye Wu, Matthew E. Piccini, and Anup K. Singh -- PNA-based

	fluorescence in situ hybridization for identification of bacteria in clinical samples/Mustafa Fazli ... [and four others].
Subjects	In situ hybridization--Laboratory manuals.
Notes	Originally published: edited by K.H. Andy Choo: 1994.
	Includes bibliographical references and index.
Series	Methods in molecular biology, 1064-3745; v. 1211 Springer protocols
	Methods in molecular biology (Clifton, N.J.); v. 1211.

MicroRNAs in medicine

LCCN	2013038052
Type of material	Book
Main title	MicroRNAs in medicine/edited by Charles H. Lawrie.
Published/Produced	Hoboken, New Jersey; Wiley Blackwell, [2014] ©2014
Description	xviii, 702 p.; ill.; 26 cm
ISBN	9781118300398 (cloth; alk. paper)
	1118300394 (cloth; alk. paper)
LC classification	QP623.5.S63 M534 2014
Related names	Lawrie, Charles H., editor of compilation.
Summary	"MicroRNAs in Medicine provides an access point into the current literature on microRNA for both scientists and clinicians, with an up-to-date look at what is happening in the emerging field of microRNAs and their relevance to medicine. Each chapter is a comprehensive review, with descriptions of the latest microRNA research written by international leaders in their field. Opening with an introduction to what microRNAs are and how they function, the book goes on to explore the role of microRNAs in normal physiological functions, infectious diseases, non-infectious diseases, cancer, circulating microRNAs as non-invasive biomarkers, and finally their potential as novel therapeutics.Including background information on the field as well as reviews of the latest research breakthroughs, MicroRNAs in Medicine is a one-stop source of information to satisfy the specialists and non-specialists alike, appealing to students,

Contents researchers, and clinicians interested in understanding the potential of microRNAs in medicine and research"--Provided by publisher.
Section IV: Hereditary and other non-infectious diseases. Chapter 28. MicroRNAs and hereditary disorders/Dr Matias Morin and Prof. Miguel Moreno-Pelayo -- Chapter 29. MicroRNAs and cardiovascular diseases/Prof. Koh Ono -- Chapter 30. MicroRNAs and diabetes/Prof. Ramono Regazzi -- Chapter 31. MicroRNAs in liver disease/Dr Patricia Munoz-Garrido, Dr Marco Marzioni, Dr Elizabeth Hijona, Dr Luis Bujanda & Dr Jesus Banales -- Chapter 32. MicroRNA regulation in multiple sclerosis/Dr Andreas Junker -- Chapter 33. The role of microRNAs in Alzheimer's disease/Dr Shahar Barbash & Prof. Hermona Soreq -- Chapter 34. Current views on the role microRNAs in psychosis/Dr Aoife Kearney, Dr Javier Bravo & Prof. Timothy Dinan. Section V: Circulating microRNAs as cellular messengers and novel biomarkers. Chapter 35. Circulating microRNAs as non-invasive biomarkers/Ass. Prof. Heidi Schwarzenbach & Prof. Klaus Pantel -- Chapter 36. Circulating microRNAs as cellular messengers/Prof. Kasey Vickers -- Chapter 37. Release of microRNA-containing vesicles can stimulate angiogenesis and metastasis in renal carcinoma/Dr Federica Collino, Dr Cristina Grange & Prof. Giovanni Camussi -- Section VI: Therapeutic uses of microRNAs: Current perspectives and future directions. Chapter 38. MicroRNA regulation of cancer stem cells and microRNAs as potential cancer stem cell therapeutics/Dr Can Liu & Prof. Dean Tang -- Chapter 39. Therapeutic modulation of microRNAs/Prof. Achim Aigner & Dr Hannelore Dassow -- Chapter 40. Silencing of MicroRNA-122 in primates with chronic hepatitis C virus infection/Dr Henrik Orum.

Subjects MicroRNAs.
Notes Includes bibliographical references and index.

Additional formats	Online version: MicroRNAs in medicine Hoboken, New Jersey; Wiley-Blackwell, [2013] 9781118300411 (DLC) 2013038614

Molecular mechanisms and physiology of disease; implications for epigenetics and health

LCCN	2014938027
Type of material	Book
Main title	Molecular mechanisms and physiology of disease; implications for epigenetics and health/Nilanjana Maulik, Tom Karagiannis, editors.
Published/Produced	New York, NY; Springer, [2014] ©2014
Description	xiv, 505 pages; illustrations (some color); 24 cm
ISBN	9781493907052 (hc; alk. paper) 1493907050 (hc; alk. paper)
LC classification	QH450 .M654 2014
Related names	Karagiannis, Tom, editor. Maulik, Nilanjana, editor.
Subjects	Epigenetics. Diseases--Molecular aspects. Epigenesis, Genetic--physiology. DNA Methylation--physiology. Epigenomics.
Notes	Includes bibliographical references and index.

Molecular stress physiology of plants

LCCN	2013931656
Type of material	Book
Main title	Molecular stress physiology of plants/Gyana Ranjan Rout, Anath Bandhu Das, editors.
Published/Produced	Dordrecht; New York; Springer India, [2013] ©2013
Description	xvii, 440 pages; illustrations (some color); 26 cm
Links	Publisher description http://www.loc.gov/catdir/ enhancements/fy1506/2013931656-d.html Table of contents only http://www.loc.gov/catdir/ enhancements/fy1506/2013931656-t.html Contributor biographical information http://www.loc. gov/catdir/enhancements/fy1506/2013931656-b.html
ISBN	9788132208068

	8132208064
LC classification	QK754 .M655 2013
Related names	Rout, Gyana Ranjan.
	Das, Anath Bandhu.
Contents	Abiotic and biostress intolerance in plants/Susana Redondo-Gomez -- Molecular mechanisms of stress resistance of photosynthetic machinery/Vladimir D. Kreslavski, Anna A. Zorina, Dmitry A. Los, Irina R. Fomina, Suleyman I. Allakhverdiev -- Salinity-induced genes and molecular basis of salt-tolerant strategies in mangroves/Anath Bandhu Das, Reto J. Strasser -- PSII fluorescence techniques for measurement of drought and high temperature stress signal in crop plants: protocols and applications/Marian Brestic, Marek Zivcak -- Salt tolerance in cereals: molecular mechanisms and applications/Allah Ditta -- Salt stress: a biochemical and physiological adapation of some Indian halophytes and sundarbans/Nirjhar Dasgupta, Paramita Nandy, Sauren Das -- Molecular physiology of osmotic stress in plants/Hrishikesh Upadhyaya, Lingaraj Sahoo, Sanjib Kumar Panda -- The physiology of reproduction-stage abiotic stress tolerance in cereals/Rudy Dolferus, Nicola Powell, Xuemei JI, Rudabe Ravash, Jane Edlington, Sandra Oliver, Joost Van Dongen, Behrouz Shiran -- Salicylic acid: role in plant physiology and stress tolerance/Gopal K. Sahu -- Role of calcium-mediated CBL-CIPK network in plant mineral nutrition and abiotic stress/Indu Tokas, Amita Pandey, Girdhar K. Pandey -- Isothermal calorimetry and Raman spectroscopy to study response of plants to abiotic and biotic stresses/Andrzej Skoczowski, Magdalena Troc -- Mechanism of plant tolerance in response to heavy metals/Jot Sharma, Nivedita Chakraverty -- Brassinosteroids: biosynthesis and role in growth, development, and thermotolerance responses/Geetika Sirhindi -- Submergence stress: responses and adaptation in crop plants/Chinmay Pradhan, Monalisa Mohanty -- Stress tolerance in plants; a proteomics approach/Gyana Ranjan Rout, Sunil Kumar Senapati -

	- Marker-assisted breeding for stress resistance in crop plants/Jogeswar Panigrahi, Ramya Ranjan Mishra, Alok Ranjan Sahu, Sobha Chandra Rath, Chitta Ranjan Kole -- DNA methylation-associated epigenetic changes in stress tolerance of plants/Mahmoud W. Yaish.
Subjects	Plants--Effect of stress on--Molecular aspects.
	Plants--Effect of stress on--Molecular aspects.
Notes	Includes bibliographical references.

Nutrition and physical activity in aging, obesity, and cancer

LCCN	2011499962
Type of material	Book
Meeting name	International Conference on Nutrition and Physical Activity in Aging, Obesity, and Cancer (2nd; 2011; Kyŏngju-si, Korea)
Main title	Nutrition and physical activity in aging, obesity, and cancer/issue editors, Young-Joon Surh ... [et al.].
Published/Created	Boston, Mass.; published by Blackwell Publishing on behalf of the New York Academy of Sciences, 2011.
Description	viii, 189 p.; ill., port.; 26 cm.
Links	Publisher description http://www.loc.gov/catdir/enhancements/fy1210/2011499962-d.html
ISBN	9781573318426
	1573318426
LC classification	RC953.5 .I58 2011
Related names	Surh, Young-Joon.
	New York Academy of Sciences.
Contents	Introduction to nutrition and physical activity in aging, obesity, and cancer/Young-Joon Surh et al. -- Xenohormesis mechanisms underlying chemopreventive effects of some dietary phytochemicals/Young-Joon Surh -- Dietary energy balance modulation of epithelial carcinogenesis: a role for IGF-1 receptor signaling and crosstalk/Tricia Moore, L. Allyson Checkley and John DiGiovanni -- Tocotrienols: inflammation and cancer/Kalanithi Nesaretnam and Puvaneswari Meganathan -- Neurogenic contributions made by dietary regulation to hippocampal neurogenesis/Hee Ra Park and

Jaewon Lee -- Growth hormone-STAT5 regulation of growth, hepatocellular carcinoma, and liver metabolism/Myunggi Baik, Ji Hoon Yu and Lothar Hennighausen -- Practical issues in genome-wide association studies for physical issues in genome-wide association studies for physical activity/Jaehee Kim et al. -- The growing challenge of obesity and cancer: an inflammatory issue/Alison E. Harvey, Laura M. Lashinger and Stephen D. Hursting -- Metabolic approaches to overcoming chemoresistance in ovarian cancer/Dong Hoon Suh et al. -- Obesity-induced metabolic stresses in breast and colon cancer/Mi-Kyung Sung et al. -- Avian biomodels for use as pharmaceutical bioreactors and for studying human diseases/Gwonhwa Song and Jae Yong Han -- Telomere dynamics: the influence of folate and DNA methylation/Carly J. Moores, Michael Fenech and Nathan J. O'Callaghan -- Stem cell engineering; limitation, alternatives, and insight/Jeong Mook Lim, et al -- Inhibitory mechanism of lycopene on cytokine expression in experimental pancreatitis/Hyeyoung Kim -- Genomic biomarkers and clinical outcomes of physical activity/Alberto Izzotti -- Impact of endurance and ultraendurance exercise on DNA damage/Karl-Heinz Wagner, Stefanie Reichhold and Oliver Neubauer -- Myricetin is a potent chemopreventive phytochemical in skin carcinogenesis/Nam Joo Kang et al. -- Culinary plants and their potential impact on metabolic overload/Ji Yeon Kim and Oran Kwon -- Inflammation-mediated obesity and insulin resistance as targets for nutraceuticals/Myung-Sunny Kim, Myeong Soo Lee and Dae Young Kown -- Assessing estrogen signaling aberrations in breast cancer risk using genetically engineered mouse models/Priscilla A. Furth et al. -- The role of carbon monoxide in metabolic disease/Yeonsoo Joe et al. -- Combinatorial strategies employing nutraceuticals for cancer development/Yogeshwer Shukla and Jasmine George -- Effects of physical activity on cancer prevention/Hye-Kyung Na and Sergiy Oliynyk --

Subjects	Regulation of the Keap1/Nrf2 system by chemopreventive sulforaphane: implications of posttranslational modifications/Young-Sam Keum. Diet therapy for older people--Congresses. Cancer--Diet therapy--Congresses. Reducing diets--Congresses. Exercise for older people--Congresses. Cancer--Exercise therapy--Congresses. Obesity--Exercise therapy--Congresses. Nutritional Physiological Phenomena--physiology--Congresses. Aging--Congresses. Diet Therapy--methods--Congresses. Exercise--physiology--Congresses. Neoplasms--diet therapy--Congresses. Obesity--diet therapy--Congresses.
Notes	"This volume presents manuscripts stemming from the "2nd International Conference on Nutrition and Physical Activity in Aging, Obesity, and Cancer (NAPA 2011)" held on february 16-19, 2011 in Gyeongju, south Korea." Includes bibliographical references.
Series	Annals of the New York Academy of Sciences, 0077-8923; v. 1229 Annals of the New York Academy of Sciences; v. 1229. 0077-8923

PIWI-interacting RNAs; methods and protocols

LCCN	2013950180
Type of material	Book
Main title	PIWI-interacting RNAs; methods and protocols/edited by Mikiko C. Siomi, Graduate School of Science, The University of Tokyo, Tokyo, Japan.
Published/Produced	New York; Humana Press, [2014] ©2014
Description	xi, 249 pages; illustrations (some color); 26 cm.
Links	Table of contents http://catdir.loc.gov/catdir/enhancements/fy1404/2013950180-t.html
ISBN	9781627036931 (alk. paper)
LC classification	QP623.5.S63 P52 2014
Related names	Siomi, Mikiko C., editor.

Contents Chromatin Immunoprecipitation Assay of Piwi in Drosophila/Hang Yin and Haifan Lin -- Drosophila Germline Stem Cells for In Vitro Analyses of PIWI-Mediated RNAi/Yuzo Niki, Takuya Sato, Takafumi Yamaguchi, Ayaka Saisho, Hiroshi Uetake, and Hidenori Watanabe -- RNAi and Overexpression of Genes in Ovarian Somatic Cells/Kuniaki Saito -- Making piRNAs In Vitro/Shinpei Kawaoka, Susumu Katsuma, and Yukihide Tomari -- A Framework for piRNA Cluster Manipulation/Ivan Olovnikov, Adrien Le Thomas, and Alexei A. Aravin -- Biochemical and Mass Spectrometric Analysis of 3′-End Methylation of piRNAs/Takeo Suzuki, Kenjyo Miyauchi, Yuriko Sakaguchi, and Tsutomu Suzuki -- HITS-CLIP (CLIP-Seq) for Mouse Piwi Proteins/Anastassios Vourekas and Zissimos Mourelatos -- DNA Methylation in Mouse Testes/Satomi Kuramochi-Miyagawa, Kanako Kita-Kojima, Yusuke Shiromoto, Daisuke Ito, Hirotaka Koshima, and Toru Nakano -- Analysis of Small RNA-Guided Endonuclease Activity in Endogenous Piwi Protein Complexes from Mouse Testes/Michael Reuter and Ramesh S. Pillai -- Small RNA Library Construction from Minute Biological Samples/Jessica A. Matts, Yuliya Sytnikova, Gung-wei Chirn, Gabor L. Igloi, and Nelson C. Lau -- Analysis of sDMA Modifications of PIWI Proteins/Shozo Honda, Yoriko Kirino, and Yohei Kirino -- Analyses of piRNA-Mediated Transcriptional Transposon Silencing in Drosophila: Nuclear Run-On Assay on Ovaries/Sergey Shpiz and Alla Kalmykova -- Combined RNA/DNA Fluorescence In Situ Hybridization on Whole-Mount Drosophila Ovaries/Shpiz, Sergey Lavrov, and Alla Kalmykova -- Fast and Accurate Method to Purify Small Noncoding RNAs from Drosophila Ovaries/Thomas Grentzinger and Séverine Chambeyron -- Isolation of Zebrafish Gonads for RNA Isolation/Hsin-Yi Huang and René F. Ketting -- Small RNA Library Construction for High-Throughput Sequencing/Jon McGinn and Benjamin Czech -- Analysis of Piwi-Loaded Small RNAs in

	Terahymena/Tomoko Noto, Henriette M. Kurth, and Kazufumi Mochizuki -- Effective Gene Knockdown in the Drosophila Germline by Artificial miRNA-Mimicking siRNAs/Hailong Wang, Haidong Huang, and Dahua Chen -- Isolation and Bioinformatic Analyses of Small RNAs Interacting with Germ Cell-Specific Argonaute in Rice/Reina Komiya and Ken-Ichi Nonomura.
Subjects	Small interfering RNA--Laboratory manuals.
	Piwi genes--Laboratory manuals.
	RNA, Small Interfering--Laboratory Manuals.
	Models, Animal--Laboratory Manuals.
	Piwi genes.
	Small interfering RNA.
Form/Genre	Laboratory Manuals.
	Handbooks, manuals, etc.
Notes	Includes bibliographical references and index.
Series	Methods in molecular biology, 1064-3745; 1093
	Springer protocols
	Methods in molecular biology (Clifton, N.J.); v. 1093.
	1064-3745
	Springer protocols (Series) 1949-2448

Plant cell culture protocols

LCCN	2012936126
Type of material	Book
Main title	Plant cell culture protocols/edited by Victor M. Loyola-Vargas, Neftalí Ochoa-Alejo.
Edition	3rd ed.
Published/Created	New York; Springer; Humana Press, c2012.
Description	xiv, 430 p.; ill. (some col.); 25 cm.
ISBN	9781617798177 (alk. paper)
	1617798177 (alk. paper)
	9781617798184 (e-ISBN)
	1617798185 (e-ISBN)
LC classification	QK725 .P5535 2012
	QH506 .M45 v.877
Related names	Loyola-Vargas, Victor M.
	Ochoa-Alejo, Neftalí.
Contents	Introduction to plant cell culture; the future ahead/Victor M. Loyola-Vargas and Neftali Ochoa-

Alejo -- History of plant tissue culture/Trevor Thorpe -- Callus, cuspension Culture, and Hairy Roots. Induction, maintenance and characterization/Rosa M. Galaz-Avalos ... [et al.] -- Growth measurements; estimation of cell division and cell expansion/Gregorio Godoy-Hernandez and Felipe A. Vazquez-Flota -- Measurement of cell viability/Lizbeth A. Castro-Concha, Rosa Maria Escobedo, and Maria de Lourdes Miranda-Ham -- Pathogen and biological contamination management in plant tissue culture; phytopathogens, vitro pathogens, and vitro pests/Alan C. Cassells -- Cryopreservation of embryogenic cell suspensions by encapsulation-vitrification and encapsulation-dehydration/Zhenfang Yin ... [et al.] -- Study of in vitro development in plants; general approaches and photography/Edward C. Yeung -- Use of statistics in plant biotechnology/Michael E. Compton -- Tissue culture methods for the clonal propagation and genetic improvement of spanish red cedar (Cedrela odorata)/Yuri Pena-Ramirez ... [et al.] -- Micropropagation of banana/Yıldız Aka Kacar and Ben Faber -- Liquid in vitro culture for the propagation of Arundo donax/Miguel Angel Herrera-Alamillo and Manuel L. Robert -- Production of haploids and doubled haploids in maize/Vanessa Prigge and Albrecht E. Melchinger -- Maize somatic embryogenesis; recent features to improve plant regeneration/Veronica Garrocho-Villegas, Maria Teresa de Jesus-Olivera, and Estela Sanchez Quintanar -- Improved shoot regeneration from root explants using an abscisic acid-containing medium/Subramanian Paulraj and Edward C. Yeung -- Cryopreservation of shoot tips and meristems; an overview of contemporary methodologies/Erica E. Benson and Keith Harding -- Anther culture of chili pepper (Capsicum spp.)/Neftali Ochoa-Alejo -- Production of interspecific hybrids in ornamental plants/Juntaro Kato and Masahiro Mii -- Plant tissue culture of fast-growing trees for phytoremediation research/Jose Luis Couselo ... [et al.] -- Removing

	heavy metals by in vitro cultures/Maria del Socorro Santos-Diaz and Maria del Carmen Barron-Cruz -- Establishment of a sanguinarine-producing cell suspension culture of Argemone mexicana L (papaveraceae); induction of alkaloid accumulation/Felipe A. Vazquez-Flota ... [et al.] -- Epigenetics, the role of DNA methylation in tree development/Marcos Viejo ... [et al.] -- Potential roles of microRNAs in molecular breeding/Qing Liu and Yue-Qin Chen -- Determination of histone methylation in mono- and dicotyledonous plants/Geovanny I. Nic-Can and Clelia De la Pena -- Basic procedures for epigenetic analysis in plant cell and tissue culture/Jose L. Rodriguez ... [et al.] -- Plant tissue culture and molecular markers/Maria Tamayo-Ordonez ... [et al.] -- Biolistic- and Agrobacterium-mediated transformation protocols for wheat/Cecilia Tamas-Nyitrai, Huw D. Jones, and Laszlo Tamas -- Improved genetic transformation of cork oak (Quercus suber L.)/Ruben Alvarez-Fernandez and Ricardo-Javier Ordas -- Organelle transformation/Anjanabha Bhattacharya ... [et al.] -- Appendix A Components of the culture media/Victor M. Loyola-Vargas -- Appendix B Plant biotechnology and tissue culture resources in the internet/Victor M. Loyola-Vargas.
Subjects	Plant cell culture--Laboratory manuals.
	Cell Culture Techniques--Laboratory Manuals.
	Plant Cells--Laboratory Manuals.
	Cells, Cultured--Laboratory Manuals.
Notes	Includes bibliographical references and index.
Series	Methods in molecular biology, 1064-3745; 877
	Springer protocols
	Methods in molecular biology (Clifton, N.J.); v. 877.
	1064-3745
	Springer protocols.

Plant epigenetics and epigenomics; methods and protocols
LCCN	2013958314
Type of material	Book

Main title	Plant epigenetics and epigenomics; methods and protocols/edited by Charles Spillane, Genetics & Biotechnology Lab, Plant & Agrosciences Centre (PABC), School of Natural Sciences, National University of Ireland, Galway (NUI Galway), Ireland, Peter C. McKeown, Genetics & Biotechnology Lab, Plant & Agrosciences Centre (PABC), School of Natural Sciences, National University of Ireland, Galway (NUI Galway), Ireland.
Published/Produced	New York; Humana Press, [2014] ©2014.
Description	x, 245 pages; illustrations (some color); 27 cm.
Links	Tabel of contents http://www.loc.gov/catdir/enhancements/fy1405/2013958314-t.html Inhaltsverzeichnis. http://bvbr.bib-bvb.de:8991/F?func=service&doc_library=BVB01&local_base=BVB01&doc_number=026996273&line_number=0001&func_code=DB_RECORDS&service_type=MEDIA
ISBN	9781627037723 (alk. paper) 1627037721 (alk. paper)
LC classification	QK981 .P49 2014
Related names	Spillane, Charlie, editor of compilation. McKeown, Peter C., editor of compilation.
Abstract	Many fundamental discoveries concerning epigenetics and the elucidation of mechanisms of epigenetic regulation have developed from studies performed in plants. In Plant Epigenetics and Epigenomics: Methods and Protocols, leading scientists in the epigenetics field describe comprehensive techniques that have been developed to understand the plant epigenetic landscape. These include recently developed methods and techniques for analysis of epigenetically regulated traits, such as flowering time, transposon activation, genomic imprinting and genome dosage effects. Written in the highly successful Methods in Molecular Biology series format, chapters include introductions to their respective topics, lists of the necessary materials and reagents, step-by-step, readily reproducible laboratory protocols, and key tips on troubleshooting and

Contents

avoidance of known pitfalls. Authoritative and practical, Plant Epigenetics and Epigenomics: Methods and Protocols seeks to aid scientists in the further study of plant epigenetic phenomena using advanced contemporary methods.-- Source other than Library of Congress.

1. Landscaping plant epigenetics/Peter C. McKeown and Charles Spillane -- 2. The gene balance hypothesis: dosage effects in plants/James A. Birchler and Reiner A. Veitia -- 3. High-throughput RNA-seq for allelic or locus-specific expression analysis in Arabidopsis-related species, hybrids, and allotetraploids/Danny W.-K. Ng, Xiaoli Shi, Gyoungju Nah and Z. Jeffrey Chen -- 4. Inference of allele-specific expression from RNA-seq data/Paul K. Korir and Cathal Seoighe -- 5. Screening for imprinted genes using high-resolution melting analysis of PCR amplicons/Robert Day and Richard Macknight -- 6. Analysis of genomic imprinting by quantitative allele-specific expression by pyrosequencing®/Peter C. McKeown, Antoine Fort and Charles Spilane -- 7. Endosperm-specific chromatin profiling by fluorence-activated nuclei sorting and chip-on-chip/Isabelle Weinhofer and Claudia Köhler -- 8. Imaging sexual reproduction in Arabidopsis using fluorescent markers/Mathieu Ingouff -- 9. Genome-wide analysis of DNA methylation in Arabidopsis using MeDIP-chip/Sandra Cortijo, René Wardenaar, Maria Colomé-Tatché, Frank Johannes and Vincent Colot -- 10. Methylation-sensitive amplified polymorphism (MSAP) marker to investigate drought-stress response in montepulciano and sangiovese grape cultivars/Emidio Albertini and Gianpiero Marconi -- 11. Detecting histone modifications inplants/Jie Song, Bas Rutjens and Caroline Dean -- 12. Quantitatively profiling genome-wide patterns of histone modifications in Arabidopsis thaliana using a chIP-seq/Chongyuan Luo and Eric Lam -- 13. Analysis of retrotransposon activity in plants/Christopher DeFraia and R. Keith Slotkin -- 14. Detecting epigenetic effects of transposable elements

	in plants/Christian Parisod, Armel Salmon, Malika Ainouche and Marie-Angèle Grandbastieu -- 15. Detection and investigation of transitive gene silencing in plants/Leen Vermeersch, Nancy De Winne and And Depicker.
Subjects	Plant genetics.
	Epigenetics.
	Plant genomes.
	Plants--genetics--Laboratory Manuals.
	Epigenesis, Genetic--Laboratory Manuals.
	Epigenetics.
	Plant genetics.
	Plant genomes.
Form/Genre	Laboratory Manuals.
Notes	Includes bibliographical references and index.
Series	Methods in molecular biology, 1064-3745; 1112 Springer protocols
	Methods in molecular biology (Clifton, N.J.); v. 1112. 1064-3745
	Springer protocols (Series) 1949-2448

Plant epigenetics; methods and protocols

LCCN	2010922364
Type of material	Book
Main title	Plant epigenetics; methods and protocols/edited by Igor Kovalchuk, Franz J. Zemp.
Published/Created	New York; Humana Press, c2010.
Description	x, 273 p.; ill. (some col.); 26 cm.
ISBN	9781607616450 (alk. paper)
	1607616459 (alk. paper)
	9781607616467 (e-ISBN)
	1607616467 (e-ISBN)
LC classification	QK981.4 .P553 2010
Related names	Kovalchuk, Igor.
	Zemp, Franz J.
Contents	Analysis of DNA methylation in plants by bisulfite sequencing/Andrea Foerster and Ortrun Mittelsten Scheid -- Analysis of bisulfite sequencing data from plant DNA using CyMATE/Andrea M. Foerster ...[et al.] -- Analysis of locus-specific changes in methylation patterns using a COBRA (combined bisulfite restriction

analysis) assay/Alex Boyko and Igor Kovalchuk -- Detection of changes in global genome methylation using the cytosine-extension assay/Alex Boyko and Igor Kovalchuk -- In situ analysis of DNA methylation in plants/Palak Kathiria and Igor Kovalchuk -- Analysis of mutation/rearrangement frequencies and methylation patterns at a given DNA locus using restriction fragment length polymorphism/Alex Boyko and Igor Kovalchuk -- Isoschizomers and amplified fragment length polymorphism for the detection of specific cytosine methylation changes/Leonor Ruiz-García ... [et al.] -- Analysis of small RNA populations using hybridization to DNA tiling arrays/Martine Boccara ... [et al.] -- Northern blotting techniques for small RNAs/Todd Blevins -- qRT-PCR of small RNAs/Erika Varkonyi-Gasic and Roger P. Hellens -- Cloning new small RNA sequences/Yuko Tagami, Naoko Inaba, and Yuichiro Watanabe -- Genome-wide mapping of protein-DNA interaction by chromatin immunoprecipitatino and DNA microarray hybridization (ChIP-chip). Part A; ChIP-chip molecular methods/Julia J. Reimer and Franziska Turck. Part B; ChIP-chip data analysis/Ulrike Göbel, Julia Reimer, and Franziska Turck -- Metaanalysis of ChIP-chip data/Julia Engelhorn and Franziska Turck -- Chromatin immunoprecipitation protocol for histone modifications and protein-DNA binding analyses in Arabidopsis/Stéphane Pien and Ueli Grossniklaus -- cDNA libraries for virus-induced gene silencing/Andrea T. Todd, Enwu Liu, and Jonathan E. Page -- Detection and quantification of DNA strand breaks using the ROPS (random oligonucleotide primed synthesis) assay/Alex Boyko and Igor Kovalchuk -- Reporter gene-based recombination lines for studies of genome stability/Palak Kathiria and Igor Kovalchuk -- Plant transgenesis/Alicja Ziemienowicz.

Subjects Plant genetic regulation.
Epigenesis.
Notes Includes bibliographical references and index.
Series Springer protocols
Methods in molecular biology, 1064-3745; v. 631
Springer protocols.

Methods in molecular biology (Clifton, N.J.); v. 631.

Probabilistic graphical models for genetics, genomics, and postgenomics

LCCN	2013953773
Type of material	Book
Main title	Probabilistic graphical models for genetics, genomics, and postgenomics/edited by Christine Sinoquet, editor-in-chief, and Raphaël Mourad, editor.
Edition	First edition.
Published/Produced	Oxford; Oxford University Press, 2014.
Description	xxvii, 449 pages, 4 unnumbered pages of plates; illustrations (some color); 25 cm
ISBN	9780198709022 (hbk.)
	0198709021 (hbk.)
LC classification	QH438.4.S73 P76 2014
Related names	Sinoquet, Christine, editor.
	Mourad, Raphaël, editor.
Contents	Pt. I. Introduction -- Probabilistic graphical models for next-generation genomics and genetics -- Essentials to understand probabilistic graphical models; a tutorial about inference and learning -- pt. II. Gene expression -- Graphical models and multivariate analysis of microarray data -- Comparison of mixture Bayesian and mixture regression approaches to infer gene networks -- Network inference in breast cancer with Gaussian graphical models and extensions -- pt. III. Causality discovery -- Utilizing genotypic information as a prior for learning gene networks -- Bayesian causal phenotype network incorporating genetic variation and biological knowledge -- Structural equation models for studying causal phenotype networks in quantitative genetics -- pt. IV. Genetic association studies -- Modeling linkage disequilibrium and performing association studies through probabilistic graphical models; a visiting tour of recent advances -- Modeling linkage disequilibrium with decomposable graphical models -- Scoring, searching and evaluating Bayesian network models of gene-phenotype association -- Graphical modeling of biological pathways in genome-wide association studies -- Bayesian systems-based, multilevel analysis

	of associations for complex phenotypes; from interpretation to decision -- pt. V. Epigenetics -- Bayesian networks in the study of genome-wide DNA methylation -- Latent variable models for analyzing DNA methylation -- pt. VI. Detection of copy number variations -- Detection of copy number variations from array comparative genomic hybridization data using linear-chain conditional random field models -- pt. VII. Prediction of outcomes from high-dimensional genomic data -- Prediction of clinical outcomes from genome-wide data.
Subjects	Genomics--Statistical methods.
	Genetics--Statistical methods.
	Graphical modeling (Statistics)
	Computational Biology--methods.
	Models, Genetic.
	Models, Statistical.
	Genomics--methods.
	Bayes Theorem.
	Computer Simulation.
	Genetics--Mathematical models.
Notes	Includes bibliographical references and index.

Pyrosequencing; methods and protocols
LCCN	2015939912
Type of material	Book
Uniform title	Pyrosequencing protocols.
Main title	Pyrosequencing; methods and protocols/edited by Ulrich Lehmann, Medizinische Hochschule Hannover, Institute of Pathology, Hannover, Germany, Jörg Tost, Laboratory for Epigenetics and Environment, Centre National de Génotypage, CEA-Institut de Génomique, Evry, France.
Edition	Second edition.
Published/Produced	New York; Humana Press, [2015] ©2015
Description	xiv, 414 pages; illustrations (some color); 26 cm.
Links	Table of contents http://www.springer.com/us/book/9781493927142
ISBN	9781493927142 (alk. paper)

Bibliography

	1493927140 (alk. paper)
LC classification	QP624 .P97 2015
Related names	Lehmann, Ulrich (Molecular biologist), editor, author.
	Tost, Jörg, editor, author.
Summary	"The primary purpose of this volume is to demonstrate the range of applications of the Pyrosequencing technology in research and diagnostics and to provide detailed protocols. Beginning with an up-to-date overview of the biochemistry, the volume continues with quantitative analysis of genetic variation, ratio of expressed alleles at the RNA level, analysis of DNA methylation, global DNA methylation assays, specialized applications for DNA methylation analysis including loss of imprinting, single blastocyst analysis, allele-specific DNA methylation patterns, DNA methylation patterns associated with specific histone modifications. The volume further details tools and protocols for the detection of viruses and bacteria, and genetic and epigenetic analyses for forensics using Pyrosequencing. As a volume in the highly successful Methods in Molecular Biology series, chapters contain introductions to their respective topics, lists of the necessary materials and reagents, step-by-step, readily reproducible protocols and tips on troubleshooting and avoiding known pitfalls. Comprehensive and adaptable, Pyrosequencing: Methods and Protocols, Second Edition will greatly aid doctorial students, postdoctoral investigators and research scientists studying different aspects of genetics and cellular and molecular biology."--Back cover.
Contents	Part I. Introduction. The History of Pyrosequencing®/Nyrén, Pål -- PyroMark® Instruments, Chemistry and Software/Kreutz, Martin (and others) -- Software-Based Pyrogram® Evaluation/Chen, Guoli (and others) -- Quantitative Validation and Quality Control of Pyrosequencing® Assays/Lehmann, Ulrich -- Part II. Analysis of Genetic Variation. Extended KRAS and NRAS Mutation Profiling by Pyrosequencing® Assays/Jung,

Andreas -- Universal BRAF State Detection by the Pyrosequencing®-Based U-BRAFv600 Assay/Skorokhod, Alexander -- Pyrosequencing®-Based Identification of Low-Frequency Mutations Enriched through Enhance ice-COLD-PCR/How-Kit, Alexandre and Jörg Tost -- Analysis of Mutational Hotspots in Routinely Processed Bone Marrow Trephines by Pyrosequencing®/Bartels, Stephan and Ulrich Lehmann -- Analysis of Copy Number Variation by Pyrosequencing® Using Paralogous Sequences/Kringen, Marianne Kristiansen -- Prenatal Diagnosis of Chromosomal Aneuploidies by Quantitative Pyrosequencing®/Ye, Hui (and others) -- HLA-B and HLA-C Supratyping by Pyrosequencing®/Vanni, Irene (and others) -- Allele Quantification Pyrosequencing® at Designated SNP Sites to Detect Allelic Expression Imbalance and Loss-of-Heterozygosity/Kwok, Chau-To and Megan P. Hitchins -- Part III. Analysis of DNA Methylation Patterns. Quantitative DNA Methylation Analysis by Pyrosequencing®/Roessler, Jessica and Ulrich Lehmann -- Quantitative Methylation Analysis of the PCDHB Gene Cluster/Banelli, Barbara and Massimo Romani -- Assessment of Changes in Global DNA Methylation Levels by Pyrosequencing® of Repetitive Elements/Tabish, Ali M. (and others) -- Global Analysis of DNA 5-Methylcytosine Using the Luminometric Methylation Assay, LUMA/Luttropp, Karin (and others) -- Limiting Dilution Bisulfite Pyrosequencing®: a Method for Methylation Analysis of Individual DNA Molecules in a Single or a Few Cells/Hajj, Nady El (and others) -- Detection of Loss of Imprinting by Pyrosequencing®/Tabano, Silvia (and others) -- Analysis of DNA Methylation Patterns in Single Blastocysts by Pyrosequencing®/Huntriss, John (and others) -- Allele-Specific DNA Methylation Detection by Pyrosequencing®/Kristensen, Lasse Sommer (and others) -- SNP-Based Quantification of Allele-Specific DNA Methylation Patterns by Pyrosequencing®/Busato, Florence and Jörg Tost -- DNA Methylation Analysis of ChIP Products at

	Single Nucleotide Resolution by Pyrosequencing®/Moison, Céline (and others) -- Part IV. Bacterial Typing and Identification. Multiplex Pyrosequencing®: Simultaneous Genotyping Based on SNPs from Distant Genomic Regions/Dabrowski, Piotr Wojciech (and others) -- Detection of Drug-Resistant Mycobacterium tuberculosis/Engström, Anna and Pontus Jureen -- Application of Pyrosequencing® in Food Biodefense/Amoako, Kingsley Kwaku -- Forensic Analysis of Mitochondrial and Autosomal Markers Using Pyrosequencing®/Buś, Magdalena M. (and others) -- Tissue-Specific DNA Methylation Patterns in Forensic Samples Detected by Pyrosequencing®/Antunes, Joana (and others).
Subjects	DNA--Analysis--Laboratory manuals.
	Nucleotide sequence--Laboratory manuals.
	Genetics--Technique--Laboratory manuals.
	Sequence Analysis, DNA--methods--Laboratory Manuals.
	High-Throughput Nucleotide Sequencing--methods--Laboratory Manuals.
	Sequence Analysis, DNA--methods--Laboratory Manuals.
	High-Throughput Nucleotide Sequencing--methods--Laboratory Manuals.
	DNA--Analysis.
	Genetics--Technique.
	Nucleotide sequence.
	Analyse de séquence d'ADN--méthodes.
	Séquence nucléotidique.
	Séquençage nucléotidique à haut débit--méthodes.
	Méthylation de l'ADN.
	Séquençage des acides nucléiques.
Form/Genre	Laboratory Manuals.
	Laboratory manuals.
Notes	Preceded by Pyrosequencing protocols/edited by Sharon Marsh. c2007.
	Includes bibliographical references and index.
Series	Methods molecular biology, 1064-3745; 1315
	Springer protocols

Methods in molecular biology (Clifton, N.J.); v. 1315.
Springer protocols (Series) 1949-2448

Speaking of genetics; a collection of interviews
LCCN   2010026037
Type of material Book
Personal name   Gitschier, Jane.
Main title   Speaking of genetics; a collection of interviews/Jane Gitschier.
Published/Created Cold Spring Harbor, N.Y.; Cold Spring Harbor Laboratory Press, c2010.
Description   x, 259 p.; ill.; 23 cm.
ISBN   9781936113033 (pbk.; alk. paper)
                    1936113031 (pbk.; alk. paper)
LC classification QH26 .G58 2010
Related titles   PLoS genetics.
Related names   Cold Spring Harbor Laboratory. Press.
Contents   In the tradition of science/Victor Ambros -- On the track of DNA Methylation/Adrian Bird -- Willing to do the math/David Botstein -- Wonderful life/Herb Boyer -- You say you want a revolution/Pat Brown -- All about mitochondrial eve/Rebecca Cann -- Curling up with a story/Sean Carroll -- Meeting a fork in the road/Tom Cech -- Twenty paces from history/Soraya de Chadarevian -- Stable in a genome of instability/Evan Eichler -- The exception that proves the rule/Jenny Graves -- The eureka moment/Sir Alec Jeffreys -- Taken to school/John E. Jones III -- The gift of observation/Mary Lyon -- Imagine/Svante Pøøbo -- The whole side of it/Neil Risch -- Ready for her close-up/Elaine Strass -- Knight in common armor/Sir John Sulston -- Sweating the details/Jamie Thomson -- The making of a president/Shirley Tilghman -- Turning the tables/Nicholas Wade -- Off the beaten path/Spencer Wells.
Subjects   Geneticists--Interviews.
                    Genetics--History--Sources.
                    Genetics--Interview.
Notes   Includes bibliographical references and index.

Spermatogenesis; methods and protocols

LCCN	2012943832
Type of material	Book
Main title	Spermatogenesis; methods and protocols/edited by Douglas T. Carrell, Kenneth I. Aston.
Published/Created	New York; Humana Press, c2013.
Description	xvii, 554 p.; ill. (some col.); 27 cm.
ISBN	9781627030373 (alk. paper)
	1627030379 (alk. paper)
	9781627030380 (eBook)
	1627030387 (eBook)
LC classification	QL966 .S6395 2013
Related names	Carrell, Douglas T.
	Aston, Kenneth I.
Contents	Methods for sperm concentration determination/Lars Bjorndahl -- Methods of sperm vitality assessment/Sergey I. Moskovtsev and Clifford L. Librach -- Hypo-osmotic swelling test for evaluation of sperm membrane integrity/Sivakumar Ramu and Rajasingam S. Jeyendran -- Sperm morphology classification; a rational method for schemes adopted by the World Health Organization/Susan A. Rothmann ... [et al.] -- Sperm morphology assessment using strict (tygerberg) criteria/Roelof Menkveld -- Methods for direct and indirect antisperm antibody testing/Hiroaki Shibahara and Junko Koriyama -- Manual methods for sperm motility assessment/David Mortimer and Sharon T. Mortimer -- Computer-aided sperm analysis (CASA) of sperm motility and hyperactivation/David Mortimer and Sharon T. Mortimer -- Hemizona assay for assessment of sperm function/Sergio Oehninger, Mahmood Morshedi, and Daniel Franken -- Sperm penetration assay for the assessment of fertilization capacity/Kathleen Hwang and Dolores J. Lamb -- Methods for the assessment of sperm capacitation and acrosome reaction excluding the sperm penetration assay/Christopher J. De Jonge and Christopher L.R. Barratt -- Sperm DNA fragmentation analysis using the TUNEL assay/Rakesh Sharma, Jayson Masaki, and Ashok Agarwal -- Sperm DNA damage measured by comet assay/Luke Simon and Douglas T. Carrell -- Sperm

chromatin structure assay (SCSA)/Donald P. Evenson -- Sperm aneuploidy testing using fluorescence in situ hybridization/Benjamin R. Emery -- Flow cytometric methods for sperm assessment/Vanesa Robles and Felipe Martinez-Pastor -- Human y chromosome microdeletion analysis by PCR multiplex protocols identifying only clinically relevant AZF microdeletions/Peter H. Vogt and Ulrike Bender -- Sperm cryopreservation methods/Tiffany Justice and Greg Christensen -- Density gradient separation of sperm for artificial insemination/David Mortimer and Sharon T. Mortimer -- Recovery, isolation, identification, and preparation of spermatozoa from human testis/Charles H. Muller and Erin R. Pagel -- Enhancement of sperm motility using pentoxifylline and platelet-activating factor/Shilo L. Archer and William E. Roudebush -- Intracytoplasmic morphology-selected sperm injection/Luke Simon, Aaron Wilcox, and Douglas T. Carrell -- Sperm selection for ICSI using annexin V/Sonja Grunewald and Uwe Paasch -- Sperm selection for ICSI using the hyaluronic acid binding assay/Mohammad Hossein Nasr-Esfahani and Tavalaee Marziyeh -- sperm selection based on electrostatic charge/Luke Simon, Shao-qin Ge, and Douglas T. Carrell -- Sex-sorting sperm using flow cytometry/cell sorting/Duane L. Garner, K. Michael Evans, and George E. Seidel -- Assessment of spermatogenesis through staging of seminiferous tubules/Marvin L. Meistrich and Rex A. Hess -- Immunohistochemical approaches for the study of spermatogenesis/Cathryn A. Hogarth and Michael D. Griswold -- Ultrastructural analysis of testicular tissue and sperm by transmission and scanning electron microscopy/Hector E. Chemes -- Assessment of oxidative stress in sperm and semen/Anthony H. Kashou, Rakesh Sharma, and Ashok Agarwal -- Improved chemiluminescence assay for measuring antioxidant capacity of seminal plasma/Charles H. Muller, Tiffany K.Y. Lee, and Michalina A. Montano -- Methods of sperm DNA extraction for genetic and epigenetic studies/Jeanine

Griffin -- Isolating mRNA and small noncoding RNAs from human sperm/Robert J. Goodrich, Ester Anton, and Stephen A. Krawetz -- Review of genome-wide approaches to study the genetic basis for spermatogenic defects/Kenneth I. Aston and Donald F. Conrad -- Methods for the analysis of the sperm proteome/Sara de Mateo, Josep Maria Estanyol, and Rafael Oliva -- Methodology of aniline blue staining of chromatin and the assessment of the associated nuclear and cytoplasmic attributes in human sperm/Leyla Sati and Gabor Huszar -- Isolation of sperm nuclei and nuclear matrices from the mouse, and other Rodents/W. Steven Ward -- Protamine extraction and analysis of human sperm protamine 1/protamine 2 ratio using acid gel electrophoresis/Lihua Liu, Kenneth I. Aston, and Douglas T. Carrell -- Analysis of gene-specific and genome-wide sperm DNA methylation/Saher Sue Hammoud, Bradley R. Cairns, and Douglas T. Carrell -- Evaluating the localization and DNA binding complexity of histones in mature sperm/David Miller and Agnieszka Paradowska -- In vitro spermatogenesis using an organ culture technique/Tetsuhiro Yokonishi ... [et al.] -- Testicular tissue grafting and male germ cell transplantation/Jose R. Rodriguez-Sosa, Lin Tang, and Ina Dobrinski -- Transgenic modification of spermatogonial stem cells using lentiviral vectors/Christina Tenenhaus Dann -- Methods for sperm-mediated gene transfer/Marialuisa Lavitrano, Roberto Giovannoni, and Maria Grazia Cerrito -- Phenotypic sssessment of male fertility status in transgenic animal models/David M. de Kretser and Liza O'Donnell.

Subjects   Spermatogenesis--Laboratory manuals.
Spermatogenesis--Laboratory Manuals.
Notes   Includes bibliographical references and index.
Series   Methods in molecular biology, 1064-3745; 927
Springer protocols
Methods in molecular biology (Clifton, N.J.); v. 927. 1064-3745
Springer protocols. 1949-2448

The chemical biology of nucleic acids
LCCN   2010000254
Type of material   Book
Main title   The chemical biology of nucleic acids/edited by Günter Mayer.
Published/Created Chichester, UK; Wiley, 2010.
Description   xiii, 464 p.; ill.; 26 cm.
ISBN   9780470519745 (cloth)
                     0470519746 (cloth)
LC classification QP620 .C44 2010
Related names   Mayer, Günter, 1972-
Contents   Chemical synthesis of modified RNA/Claudia Höbartner and Falk Wachowius -- Expansion of the genetic alphabet in nucleic acids by creating new base pairs/Ichiro Hirao and Michiko Kimoto -- Chemical biology of DNA replication; probing DNA polymerase selectivity mechanisms with modified nucleotides/Andreas Marx -- Nucleic-acid-templated chemistry/Michael Oberhuber -- Chemical biology of peptide nucleic acids (PNA)/Peter E. Nielsen -- The interactions of small molecules with DNA and RNA/Yun Xie, Victor Tam and Yitzhak Tor -- The architectural modules of folded RNAs/V. Fritsch and Eric Westhof -- Genesis and biological applications of locked nucleic acid (LNA)/Harleen Kaur and Souvik Maiti -- Small non-coding RNA in bacteria/Sabine Brantl -- Microrna-guided gene silencing/Gunter Meister -- Nucleic acids based therapies/Britta Hoehn and John J. Rossi -- Innate immune recognition of nucleic acid/Stefan Bauer -- Light-responsive nucleic acids for the spatiotemporal control of biological processes/Alexander Heckel and Günter Mayer -- DNA methylation/Albert Jeltsch and Renata Z. Jurkowska -- Frameworks for programming RNA devices/Maung Nyan Win, Joe C. Liang and Christina D. Smolke -- RNA as a catalyst; the Diels-Alderase-ribozyme/Andres Jäschke -- Evolving an understanding of RNA function by in vitro approaches/Qing Wang and Peter J. Unrau -- The chemical biology of aptamers; synthesis and applications/Günter Mayer & Bernhard Wulffen --

Subjects	Nucleic acids as detection tools/Jeffrey C.F. Lam, Sergio Aguirre and Yingfu Li -- Bacterial riboswitch discovery and analysis/Tyler D. Ames and Ronald R. Breaker. Nucleic acids. Nucleic Acids--chemistry. Nucleic Acids--physiology. Nucleic Acids--therapeutic use.
Notes	Includes bibliographical references and index.

The zebrafish; genetics, genomics and informatics

LCCN	2011499869
Type of material	Book
Main title	The zebrafish; genetics, genomics and informatics/edited by H. William Detrich III, Monte Westerfield, Leonard I. Zon.
Edition	3rd ed.
Published/Created	Amsterdam [Netherlands]; Boston [Mass.]; Elsevier/Academic, 2011.
Description	xvii, 504 p., [32] p. of color plates; ill.; 26 cm.
ISBN	9780123748140 (hbk.) 0123748143 (hbk.)
LC classification	QL638.C94 Z433 2011
Related names	Detrich, H. William. Westerfield, Monte. Zon, Leonard I.
Contents	Pt.1 Forward and reverse genetics; Generating conditional mutations in zebrafish using gene-trap mutagenesis/Lisette A. Maddison, Jianjun Lu, Wenbiao Chen -- Tol2-mediated transgenesis, gene trapping, enhancer trapping, and the Gal4-UAS system/Gembu Abe, Maximilliano L. Suster, Koichi Kawakami -- Engineering zinc finger nucleases for targeted mutagenesis of zebrafish/Jeffry D. Sander, J.-R. Joanna Yeh, Randall T. Peterson, J. Keith Joung -- Retroviral-mediated insertional mutagenesis in zebrafish/Adam Amsterdam, Gaurav Kumar Varshney, Shawn Michael Burgess -- Genetic screens for mutations affecting adult traits and parental-effect genes/Francisco Pelegri, Mary C. Mullins -- High-throughput target-selected gene inactivation in zebrafish/Ross N.W. Kettleborough, Ewart de Bruijn,

Freek van Eeden, Edwin Cuppen, Derek L. Stemple --
Genetic suppressor screens in haploids/Xiaoying Bai,
Zhongan Yang, Hong Jiang, Shuo Lin, Leonard I. Zon
-- Transgenic zebrafish using transposable
elements/Karl J. Clark, Mark D. Urban, Kimberly J.
Skuster, Stephen C. Ekker -- Spatiotemporal control
of embryonic gene expression using caged
morpholinos/Ilya A. Shestopalov, James K. Chen --
Pt.2 Transgenesis; Advanced zebrafish transgenesis
with Tol2 and application for Cre/lox recombination
experiments/Christian Mosimann, Leonard I. Zion --
Use of phage PhiC31 integrase as a tool for zebrafish
genome manipulation/James A. Lister -- Method for
somatic cell nuclear transfer in zebrafish/Kannika
Siripattarapravat, Jose B. Cibelli -- Pt.3 The zebrafish
genome and mapping technologies; Single nucleotide
polymorphism (SNP) panels for rapid positional
cloning in zebrafish/Matthew D. Clark, Victor
Guryev, Ewart de Bruijn, Isaac J. Nijman, Masazumi
Tada, Catherine Wilson, Panos Deloukas, John H.
Postlethwait, Edwin Cuppen, Derek L. Stemple --
Molecular cytogenetic methodologies and a BAC
probe panel resource for genomic analyses in the
zebrafish/Kimberly P. Dobrinski, Kim H. Brown,
Jennifer L. Freeman, Charles Lee -- Conserved
synteny and the zebrafish genome/Julian M. Catchen,
Ingo Braasch, John H. Postlethwait -- The zon
laboratory guide to positional cloning in zebrafish/Yi
Zhou, Leonard I. Zon -- Pt.4 Informatics and
comparative genomics; Data extraction,
transformation, and dissemination through
ZFIN/Douglas G. Howe, Ken Frazer, David Fashena,
Leyla Ruzicka, Yvonne Bradford, Sridhar
Ramachandran, Barbara J. Ruef, Ceri Van Slyke,
Amy Singer, Monte Westerfield -- DNA methylation
profiling in zebrafish/Shan-Fu Wu, Haiying Zhang,
Saher Sue Hammond, Magdalena Potok, David A.
Nix, David A. Jones, Bradley R. Cairns -- Chromatin
immunoprecipitation in adult zebrafish red cells/Eirini
Trompouki, Teresa Venezia Bowman, Anthony
DiBiase, Yi Zhou, Leonard I. Zon -- Discerning

Subjects	different in vivo roles of microRNAs by experimental approaches in zebrafish/Luke Pase, Graham J. Lieschke -- Sequencing-based expression profiling in zebrafish/Jin Liang, Gabriel Renaud, Shawn M. Burgess -- Chromatin modification in zebrafish development/Jordi Cayuso Mas, Emily S. Noël, Elke A. Ober -- Pt.5 Infrastructure; Advances in zebrafish husbandry and management/Christian Lawrence -- Aquaculture and husbandry at the Zebrafish International Resource Center/Zoltán M. Varga. Zebra danio. Zebra danio--Research--Methodology. Zebrafish--genetics. Genome.
Notes	Includes bibliographical references and index.
Series	Methods in cell biology, 0091-679X; v. 104 Methods in cell biology; v. 104.

Toxicology and epigenetics

LCCN	2012010175
Type of material	Book
Main title	Toxicology and epigenetics/editor, Saura C. Sahu.
Published/Created	Chichester, West Sussex, U.K.; Hoboken, N.J.; John Wiley & Sons, 2012.
Description	xxix, 658 p.; ill.; 26 cm.
ISBN	9781119976097 (cloth)
LC classification	RA1224.3 .T698 2012
Related names	Sahu, Saura C.
Contents	Environment, Epigenetics and Diseases/Robert Y.S. Cheng and Wan-yee Tang -- DNA Methylation and Toxicogenomics/Deepti Dileep Deobagkar -- Chromatin at the Intersection of Disease and Therapy/Delphine Quénet, Marcin Walkiewicz and Yamini Dalal -- Molecular Epigenetic Changes Caused by Environmental Pollutants/Solange S. Lewis, Gregory J. Weber, Jennifer L. Freeman and Maria S. Sepulveda -- Epigenetic mediation of environmental exposures to Polycyclic Aromatic Hydrocarbons/Bekim Sadikovic and David I. Rodenhiser -- Epigenomic Actions of Environmental Arsenicals/Paul L. Severson & Bernard W. Futscher --

Arsenic-induced Changes to the Epigenome/Kathryn A. Bailey and Rebecca C. Fry -- Environmental Epigenetics, Asthma and Allergy; Our environment's Molecular Footprints/Stephanie Lovinsky-Desir and Rachel L. Miller -- miRNAs in Human Prostate Cancer/Ernest K. Amankwah and Jong Y. Park -- Environment, Epigenetics and Cardiovascular Health/Sanjukta Ghosh and Andrea Baccarelli -- Toxicology, Epigenetics and Autoimmunity/Craig A. Cooney and Kathleen M. Gilbert -- Toxicoepigenomics in Lupus/Donna Ray and Bruce C. Richardson -- Ocular Epigenomics; Potential Sites of Environmental Impact in Development and Disease/Kenneth P. Mitton -- Nuclear RNA Silencing and Related Phenomena in Animals/Radek Malik and Petr Svoboda -- Epigenetic Biomarkers in Cancer Detection and Diagnosis/Ashley G. Rivenbark and William B. Coleman -- Epigenetic Histone Changes in the Toxicologic mode of action of Arsenic/John F. Reichard and Alvaro Puga -- Irreversible Effect of Diethylstilbestrol on Reproductive Organs and Current Approach for Epigenetic Effects of Endocrine Disrupting Chemicals/Shinichi Miyagawa, Ryohei Yatsu, Tamotsu Sudo, Katsuhide Igarashi, Jun Kanno, and Taisen Iguchi -- Epigenomics; Impact for Drug Safety Sciences/Harri Lempiäinen, Raphaelle Luisier, Arne Muller, Philippe Marc, David Heard, Federico Bolognani, Pierre Moulin, Philippe Couttet, Olivier Grenet, Jennifer Marlowe, Jonathan Moggs, and Rémi Terranova -- Archival Toxicoepigenetics; Molecular Analysis of Modified DNA from Preserved Tissues in Toxicology Studies/B. Alex Merrick -- Nanoparticles and Toxicoepigenomics/Manasi P. Jain, Angela O. Choi, and Dusica Maysinger -- Methods of Global Epigenomic Profiling/Michael W.Y. Chan, Zhengang Peng, Jennifer R. Chao, Yingwei Li, Matthew T. Zuzolo, and Huey-Jen L. Lin -- Transcriptomics; Applications in Epigenetic Toxicology/Pius Joseph -- Carcinogenic metals alter histone tail modifications/Yana Chervona and Max Costa -- Prediction of epigenetic and stochastic gene

	expression profiles of late effects after radiation exposure/Yoko Hirabayashi and Tohru Inoue -- Modulation of developmentally regulated gene expression programs through targeting of polycomb and trithorax group proteins/Marjorie Brand and F. Jeffrey Dilworth -- Chromatin Insulators and Epigenetic Inheritance in Health and Disease/Jingping Yang and Victor G. Corces -- Bioinformatics for High-Throughput Toxico-epigenomics Studies/Maureen A. Sartor, Dana C. Dolinoy, Laura S. Rozek, and Gilbert S. Omenn -- Computational Methods in Toxicoepigenomics/Joo Chuan Tong -- Databases and Tools for Computational Epigenomics/V. Umashankar and S. Gurunathan -- Interface of Epigenetics and Carcinogenic Risk Assessment/Paul Nioi -- Epigenetic Modifications in Chemical Carcinogenesis/Igor P. Pogribny, Igor Koturbash, and Frederick A. Beland -- Application of Cancer Toxicoepigenomics in Identifying High-Risk Populations/Mukesh Verma and Krishna K Banaudha.
Subjects	Genetic toxicology.
	Environmental toxicology.
	Epigenetics.
Notes	Includes bibliographical references and index.

RELATED NOVA PUBLICATIONS

DNA METHYLATION: PRINCIPLES, MECHANISMS AND CHALLENGES

Editors: Tatiana V. Tatarinova and Gaurav Sablok
Division of Mathematics and Statistics, University of Glamorgan, Pontypridd, United Kingdom

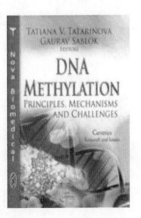

ISBN: 978-1-62417-128-4   **Publication Date:** 2013

DNA methylation is a cryptic phenomenon that invokes the methylation of the cytosines in nuclear DNA and is responsible for a wide variety of essential processes, starting from cellular differentiation (embryogenesis), transposon silencing, miRNA dependent methylation and gene regulation. This book presents an overview of different aspects of DNA methylation with a focus on its basic principles and mechanisms and gene silencing. Also discussed, is the role of DNA methylation in plants; epigenetic control of circadian clock operation; photoperiodic flowering; and DNA methylation in cancer and its role in multiple sclerosis.

GLOBAL DNA METHYLATION IN OBESITY, DIABETES AND CARDIOVASCULAR DISEASES AND THE INFLUENCE OF ENVIRONMENTAL FACTORS[*]

M. Garcia-Lacarte[1], F.I. Milagro[1,2] and J.A. Martinez[1,2]

[1]Department of Nutrition, Food Science and Physiology and the Centre for Nutrition Research, University of Navarra, Pamplona, Spain.

[2]Instituto de Salud Carlos III, CIBER Fisiología Obesidad y Nutrición (CIBERobn), Madrid, Spain.

As the prevalence of obesity and related diseases such as cardiovascular disease and type 2 diabetes is increasing worldwide, different strategies and novel approaches are being investigated to manage these pathological conditions. In this context, dynamic changes in DNA as a consequence of environmental interactions have gained interest in understanding these diseases. Indeed, several studies have demonstrated that the epigenome is affected by external factors, such as diet, physical activity, stress, or exposure to chemical carcinogens. Global methylation status is commonly used as a surrogate measure of overall methylation changes. Actually, aberrations in DNA methylation are often related to disease. In this sense, global DNA methylation pattern changes, measured in *LINE-1* and *Alu* sequences, have been associated to body weight regulation and the onset of obesity, cardiovascular diseases and type 2 diabetes, and could be used as possible diagnostic and prognostic biomarkers. In relation to cardiovascular diseases, *LINE-1* is the principal sequence that trends to hyper or hypomethylation depending on the tissue analysed. In type 2 diabetes, changes in *LINE-1* methylation levels have been

---

[*] The full version of this chapter can be found in *Gene-Environment Interactions and Human Diseases*, edited by Lu Qi, published by Nova Science Publishers, Inc, New York, 2015.

reported in peripheral blood mononuclear cells (PBMC). Focusing on obesity, several studies have reported changes in the methylation status of both, *LINE-1* and *Alu* sequences. For the near future, the principal challenges in nutriepigenetics are the large number of variables, intermediate markers and measurements to be considered, as well as the dynamic nature of nutrients and the limited outcome information.

THE LINK BETWEEN *MTHFR C677T* POLYMORPHISM, FOLATE METABOLISM AND GLOBAL DNA METHYLATION: A LITERATURE REVIEW[*]

Antonella Agodi[†], Annalisa Quattrocchi, Andrea Maugeri and Martina Barchitta
Department of Medical and Surgical Sciences and Advanced Technologies "GF Ingrassia," University of Catania, Catania, Italy

DNA methylation is an epigenetic phenomenon that affects the regulation of gene expression and genome integrity. Folate, providing methyl groups for DNA methylation, plays a key role in the maintenance of genomic stability. Accordingly, folate-mediated one-carbon metabolism is linked to DNA methylation status but it is also influenced by genetic polymorphisms. Methylenetetrahydrofolate reductase (*MTHFR*), which is involved in the supply of the methylation group, is an enzyme necessary for the folate metabolic pathway and is considered to result in hypermethylation of genomic DNA. Particularly, the *MTHFR C677T* polymorphism results in an alanine (C)-to-valine (T) substitution and makes the enzyme less active. This literature review focuses on the recent evidence-based reports on the relationship between *MTHFR C677T* polymorphism, folate metabolism and global DNA methylation in various physiological or pathological conditions.

---

[*] The full version of this chapter can be found in *Methylenetetrahydrofolate Reductase (MTHFR) in Health and Disease*, edited by Roger Evans, published by Nova Science Publishers, Inc, New York, 2015.
[†] Corresponding author: Antonella Agodi, Department of Medical and Surgical Sciences and Advanced Technologies "GF Ingrassia," University of Catania, via S. Sofia, 87 – 95121 Catania, Italy. Phone/Fax: +390953782183; e-mail: agodia@unict.it.

Medline database was searched to select articles, published before January 2015, investigating the link between *MTHFR C677T* polymorphism and global DNA methylation. Information about Author's last name, year of publication, country where the study was performed, experimental methods and outcome were collected in a standard format.

The number of searched articles was limited (24) and selected studies were highly heterogeneous regarding to study design, sample source, experimental methods and outcome. Overall, no significant relationship between *MTHFR C677T* polymorphism and DNA methylation was reported, although only one study showed a significant difference in methyl group acceptance capacity between carriers of the *MTHFR677T/T* genotype and carriers of the wild-type *MTHFR677C/C* genotype. The enhanced methyl group acceptance capacity in *MTHFR677T/T* carriers indicated genomic hypomethylation status compared to carriers of *MTHFR677C/C* genotype. In addition, a further study revealed that reduction in DNA methylation levels was significantly associated with the T allele only in those subjects with low plasma folate concentrations.

Evidence from this review, conducted on a small number of subjects, did not support the hypothesis that decreased global DNA methylation status is directly associated with *MTHFR C677T* polymorphism. However, if associated with low plasma folate, the *MTHFR 677TT* homozygous mutant genotype appears to be crucial in determining the lowest DNA methylation levels. Thus, only the interaction between *MTHFR* polymorphisms and folate deficiency may play a potential role in the complex mechanism of global DNA hypomethylation.

THE ROLE OF DNA METHYLATION IN THE GENETICS AND EPIGENETICS OF MULTIPLE MYELOMA*

Hiroshi Yasui[†,1]*, Tadao Ishida*[3] *and Kohzoh Imai*[2]

[1]Center for Antibody and Vaccine, Research Hospital,
[2]The Institute of Medical Science, The University of Tokyo, Tokyo, Japan
[3]Department of Gastroenterology, Rheumatology,
and Clinical Immunology, Sapporo Medical University, Sapporo, Japan

Multiple myeloma (MM) arises through an accumulation of multiple genetic and epigenetic changes, which play a significant role in tumorigenesis and tumor development. DNA methylation is often found in cancers including MM at the 5-carbon on cytosine residues within CpG islands of genes whose products are associated with the promoter regions of protein-coding genes. This methylation is an epigenetic alteration that leads to heritable changes in gene expression through the recruitment of histone deacetylases and histone methyltransferases. We and other researchers have reported the association of global and regional DNA methylation status with MM. Global DNA hypomethylation is the predominant early change during plasma cell oncogenesis from monoclonal gammopathy of undetermined significance to MM, while regional DNA hypermethylation occurs in tumor relapse and during disease progression. Thus, DNA methylation could be a useful biomarker of MM tumorigenesis and progression. In the current review, we discuss the role of DNA methylation changes; their potential application as epigenetic biomarkers to facilitate risk assessment, diagnosis, prediction of prognosis, and sensitivity to treatment; and epigenetic therapy in MM.

---

* The full version of this chapter can be found in *Multiple Myeloma: Risk Factors, Diagnosis and Treatments*, edited by Steve Holt, published by Nova Science Publishers, Inc, New York, 2014.
† Corresponding author: Hiroshi Yasui, M.D., Ph.D. Center for Antibody and Vaccine, Research Hospital, The Institute of Medical Science, The University of Tokyo; 4-6-1 Shirokanedai, Minato-ku, Tokyo 108-8639, Japan; TEL: +81-3-3443-8111; FAX: +81-3-6409-2413; E-mail: hiroyasu@ims.u-tokyo.ac.jp

DETECTING DNA METHYLATION FOR CANCER DIAGNOSTICS AND PROGNOSTICS[*]

Eugene J. H. Wee[†], Muhammad J. A. Shiddiky and Matt Trau
Australian Institute for Bioengineering and Nanotechnology,
The University of Queensland, St Lucia, Qld, Australia

DNA methylation is an epigenetic process that has roles in many normal cellular processes and dysregulation of which can result in diseases such as cancer. The most well studied form of mammalian DNA methylation is the addition of a methyl group to the number 5 carbon of cytosine in CpG dinucleotides.

Gene regulatory elements, such as gene promoters and enhancers associated with dense CpGs, are sensitive to DNA methylation silencing. Multiple studies have shown that promoter methylation profiling of cancer genes may be useful as biomarkers of cancer. DNA methylation information has the potential to provide information on a patient's cancer subtype, treatment response and prognosis. Current DNA methylation detection techniques can be broadly grouped into sodium bisulfite, restriction enzyme and methylated DNA enrichment based techniques. However, techniques for detection DNA methylation have largely been tailored to research purposes and may not be well suited for routine clinical use.

In this chapter, some of these common DNA methylation detection methods are reviewed for their suitability in diagnostics. Also discussed are ways how some of these techniques may be or have been adapted for clinical and point-of-care applications. Emerging techniques that have evolved from classical research methods are also introduced.

---

[*] The full version of this chapter can be found in *Microfluidics, Nanotechnology and Disease Biomarkers for Personalized Medicine Applications*, edited by Muhammad J. A. Shiddiky, Eugene J. H. Wee, Sakandar Rauf and Matt Trau, published by Nova Science Publishers, Inc, New York, 2013.
[†] E-mail: j.wee@uq.edu.au; Tel: +61 07 33464176.

INDEX

#

2,3,7,8-tetrachlorodibenzo-p-dioxin, 4, 5
3D DNA, 42, 43, 44, 45, 46
4meC (N4-methylcytosine), 30, 35, 59
5'-carboxycytosine (5caC), 29, 30, 32, 33, 34, 37, 45, 46, 53, 57, 83
5'-formylcytosine (5fC), 29, 30, 32, 33, 34, 37, 41, 43, 45, 46, 47, 53, 57, 83
5'-hydroxymethylcytosine (5hmC), 29, 30, 32, 33, 34, 37, 41, 43, 45, 46, 47, 50, 52, 53, 55, 57, 63, 67, 78, 83, 147
5'-methylcytosine (5meC), 29, 30, 32, 33, 34, 37, 41, 43, 45, 46, 47, 50, 52, 53, 57, 59

A

abuse, 150, 151
access, 52, 58, 107, 160
accessibility, 120
acetylation, 7, 22, 35, 36, 44, 52, 76, 82, 85, 121, 131, 137
acid, 48, 117, 137, 141, 163, 169, 182, 184
acidic, 53
acrosome, 181
activating transcription factor 4 (ATF4), 30, 54, 55, 83
activation induced deaminase (AID), 30, 34, 37, 70
active demethylation, 29, 34, 55, 58, 69, 71
acute myeloid leukemia, 102, 109
acute stress, 84
adaptation, 113, 142, 143, 164
adenocarcinoma, 103, 119
adenovirus, 98, 99, 100, 109
adipocyte, 81
adiposity, 13, 19, 25
adolescents, 22
ADP, 100
adulthood, 13, 18, 20
adults, 10, 64
adverse effects, 4, 6, 8, 9, 14, 16
AFM, 147
age, 8, 22, 24, 56, 59, 142, 179
Agrobacterium, 170
airway inflammation, 23
alanine, 193
alcohol consumption, 23
alcohols, 47
aldehyde, 34, 47
algorithm, 48
allele, 32, 38, 56, 92, 93, 94, 95, 96, 107, 172, 177, 194
allergens, 111
allergy, 23
allometry, 142

altered, 4, 5, 7, 8, 9, 10, 12, 13, 14, 55, 86, 96, 106, 127
alters, 6, 12, 16, 19, 22, 23, 75, 76, 105
alzheimer disease, 33
amines, 47
amino, 26, 27, 48, 51, 53
amino acid, 26, 27, 48, 51, 53
amphibia, 57, 86
aneuploidy, 181
angiogenesis, 161
aniline, 182
antibody, 102, 181
anti-cancer, 91, 92, 98
antigen, 13, 31, 51, 89, 94, 104, 108
antigen-presenting cell, 14
antioxidant, 123, 182
antisense, 116
antisense RNA, 116
anxiety, 13, 20
APC, 127
apolipoprotein B mRNA editing enzyme, catalytic polypeptide-like (APOBEC), 30, 34, 70
apoptosis, 100, 147
arabidopsis, 41, 51, 56, 58, 84, 108, 126, 172, 174
Arabidopsis thaliana, 58, 172
Arg, 30, 48
arginine, 30, 46, 47, 48, 49, 51, 79
aromatic rings, 46
arsenic, 8, 19, 21, 22
arthritis, 130
arthrobacter luteus restriction endonuclease (Alu), 30, 59, 192
aryl hydrocarbon receptor, 4
assessment, 111, 123, 125, 147, 155, 181
asthma, 23, 128, 135
asymmetry, 134
atherosclerosis, 130
atoms, 45
ATP, 121
attachment, 142
autism, 130, 140
autoantigens, 116
autoimmune disease, 136
autoimmunity, 149
avoidance, 171

B

BAC, 186
bacteria, 32, 59, 113, 136, 160, 177, 184
bacterial DNA, 59
bacterial genome, 35, 59, 113, 155
bacterial pathogens, 155
bacterium, 155
base, 22, 27, 29, 32, 34, 37, 42, 43, 44, 58, 69, 70, 75, 83, 88, 171, 184
base pair, 42, 43, 44, 83, 184
behavioral change, 22
behaviors, 7, 142
bending, 43, 45, 48, 52
benzene, 123
B-form, 42
bilateral, 60, 62
bioavailability, 13
biochemistry, 137, 177
bioinformatics, 149
biological activities, 33, 49
biological processes, 29, 60, 140, 184
biological systems, 15
biomarkers, 21, 91, 135, 151, 159, 161, 165, 192, 195, 196
biosynthesis, 163
biotechnology, 111, 169
biotic, 163
biotin, 160
bipolar disorder, 140
birth weight, 20, 21, 23
bisphenol, 4, 5, 22
bisphenol A, 4, 5, 17, 22, 27
black carbon, 10
bladder cancer, 66
blood, 7, 8, 10, 12, 13, 14, 21, 22, 24
body composition, 141, 143
body size, 143
body weight, 192
bonding, 45, 48, 52, 146
bonds, 45
BPA, 4

Index

brain, 2, 4, 7, 8, 9, 13, 14, 15, 18, 19, 22, 25, 26, 27, 33, 56, 57, 67, 72, 142, 150
brain damage, 14
brain tumor, 9, 25
breast cancer, 65, 67, 93, 100, 140, 165, 175
breeding, 164, 170
broad-complex, tramtrack and bric a brac (BTB Domain), 30, 32
bromination, 75
budding, 120

C

cadmium, 6, 7, 18, 20, 21, 24
calcium, 163
calorimetry, 163
cancer, 33, 38, 40, 49, 51, 57, 58, 65, 66, 68, 69, 74, 85, 86, 91, 92, 93, 94, 96, 97, 100, 101, 102, 103, 107, 108, 109, 114, 115, 117, 118, 127, 130, 131, 132, 134, 136, 139, 140, 145, 161, 164, 192, 195, 196
cancer cells, 40, 58, 66, 68, 100, 103, 107, 118
cancer progression, 102, 119
cancer stem cells, 137, 161
cancer therapy, 115
cannabis, 150
carbohydrate, 13
carbon, 10, 14, 32, 33, 35, 44, 45, 46, 165, 193, 195, 196
carbon monoxide, 14, 165
carbon-hydrogen bond, 46
carboxyl, 34, 86
carcinogenesis, 51, 55, 117, 139, 164
carcinoma, 66, 161
cardiac structure, 19
cardiovascular disease, 17, 25, 161, 192
cardiovascular disorders, 9
cascades, 92
Caspase-8, 85
catalyst, 184
cation, 46, 48, 77
cation – π interaction, 46
C-C, 76, 77

CCAAT-enhancer-binding proteins (C/EBPs), 30
CCCTC-binding factor (CTCF), 30, 49, 81, 91, 95, 96, 97, 98, 99, 101, 104, 105, 106, 107, 108
CD8+, 7, 8
cDNA, 98, 152, 174
cell biology, 97, 187
cell culture, 168, 169, 170
cell cycle, 34, 38, 133
cell death, 33, 49
cell differentiation, vii, 2, 3, 123, 136
cell division, 3, 139, 169
cell fate, 83, 108, 127, 132
cell line, 80, 96, 97, 98, 99, 100, 105, 119
central nervous system, 9, 103
cerebellum, 41
cerebral palsy, 7
cervical cancer, 117, 118
cervix, 117, 119
CG base pair, 42
CGC, 52
CG-medium, 41
CG-poor, 41
CG-rich, 41
Chad, 126
challenges, 21, 111, 124, 193
chemical, 2, 3, 4, 14, 28, 33, 36, 45, 47, 52, 53, 58, 60, 62, 134, 183, 184, 192
chemical bonds, 45, 62
chemical characteristics, 45, 53
chemical reactions, 46, 58
chemiluminescence, 182
chemoprevention, 23, 135
chemotherapeutic agent, 92
chemotherapy, 127
chicken, 79
childhood, 9, 15, 25, 93, 130, 142
children, 24
chimeric, 53, 126
China, 91
chloroplast, 41, 70
choline, 27
Christians, 79
chromatid, 146

Index

chromatin, 24, 25, 38, 42, 47, 51, 52, 62, 64, 67, 69, 72, 74, 78, 79, 80, 81, 82, 90, 91, 96, 97, 98, 101, 106, 107, 120, 122, 133, 136, 137, 138, 139, 146, 148, 152, 167, 172, 174, 181, 186, 187
chromatography, 79, 147
chromosome, 2, 3, 11, 12, 38, 64, 92, 95, 152, 181
chronic diseases, 128
chronic obstructive pulmonary disease, 23, 129
cigarette smoke, 2, 6, 14, 15, 21, 22, 26, 85
classes, 4, 73
classification, 37, 111, 113, 114, 115, 117, 118, 120, 122, 124, 125, 127, 128, 129, 130, 131, 133, 134, 136, 137, 139, 141, 144, 146, 148, 149, 150, 151, 153, 155, 156, 158, 160, 162, 163, 164, 167, 168, 171, 173, 175, 176, 180, 181, 183, 185, 187
cleavage, 46, 70
cloning, 125, 149, 186
CNS, 92, 93
coding, 55, 65, 84, 93, 132, 184, 195
collaboration, 115
colon, 40, 66, 80, 85, 95, 98, 100, 101, 104, 108, 165
colon cancer, 40, 66, 80, 95, 98, 100, 101, 108, 165
color, 113, 131, 150, 159, 162, 166, 171, 175, 176, 185
colorectal cancer, 58, 84, 86, 93, 94, 102, 104, 105, 106, 109, 140
comparative analysis, 64
competition, 95
compilation, 160, 171
complementary DNA, 45, 53
complexity, 43, 47, 73, 89, 183
complications, 28, 136
composition, 147, 155
compounds, 4, 8, 14, 17, 19, 53
Concise, 108
configuration, 96, 107
conformational changes, 43
Congress, 172
consensus, 17
conservation, 86
constituents, 134
construction, 10, 16, 120, 144
consumption, 19
contamination, 169
control group, 9
controversial, 4
COOH, 34, 58
coordination, 91
coronary heart disease, 17
correlation, 58, 116, 121
covalent bond, 45
CpG (cytosine-phosphate-guanine nucleotide), vii, 1, 2, 6, 9, 11, 12, 14, 15, 17, 18, 20, 21, 27, 30, 31, 32, 33, 38, 40, 41, 43, 44, 52, 55, 56, 62, 63, 64, 65, 68, 72, 73, 74, 76, 78, 79, 80, 81, 82, 83, 87, 88, 95, 96, 137, 195, 196
CpG content, 41, 44, 52, 62
CpG intensity, 41
CpG islands, 1, 2, 6, 9, 11, 12, 14, 18, 41, 56, 64, 65, 74, 80, 95, 137, 195
CpG sites, 1, 3, 15, 27
crop, 150, 163
cryopreservation, 182
cryptorchidism, 154
CTCF, 30, 49, 81, 91, 95, 96, 97, 98, 99, 101, 104, 105, 106, 107, 108
cultivars, 172
culture, 118, 123, 145, 154, 169, 183
culture media, 170
CXXC domain, 49
cyclin-dependent kinase inhibitor (CDKN), 8, 30
cytochrome, 123
cytometry, 182
cytoplasm, 97
cytosine, 1, 2, 26, 27, 29, 30, 32, 33, 34, 35, 36, 37, 41, 42, 43, 44, 45, 46, 47, 48, 53, 56, 57, 58, 59, 60, 61, 62, 68, 69, 72, 73, 75, 76, 83, 86, 88, 89, 147, 173, 195, 196
cytotoxicity, 100, 123

Index

D

DAPK1 (death associated protein kinase 1), 30, 33, 66, 89
data analysis, 114, 174
database, 194
de novo, 1, 3, 13, 16, 24, 34, 68, 99, 114
deacetylation, 52
deamination, 34, 69, 77
defects, 2, 13, 26, 182
deficiency, 13, 26, 33, 68, 194
deformability, 78
degradation, 118, 132
DEHP, 5, 6, 22
dehydration, 169
denaturation, 126
Denmark, 158
deoxyribonucleic acid, 69
Department of Defense, 101
depression, 24
derivatives, 37
detectable, 40, 90
detection, 41, 86, 89, 91, 103, 118, 121, 147, 159, 174, 177, 184, 196
detection techniques, 196
di(2-ethylhexyl) phthalate, 5, 6
diabetes, 18, 19, 25, 64, 92, 128, 136, 161, 192
diesel exhaust, viii, 2, 9, 11, 12, 18, 20, 21, 23, 26, 27
diet, 2, 12, 13, 19, 20, 22, 27, 59, 104, 166, 192
dietary intake, 6, 13, 24
differentially methylated regions (DMRs), 5, 6, 8, 30, 38, 56, 71, 95, 96, 97, 106, 107
DIGE, 116
digestion, 153
dimorphism, 143
dioxin, 26
disease model, 149, 150
disease progression, 195
diseases, 2, 3, 13, 25, 35, 128, 129, 130, 131, 137, 140, 150, 161, 165, 192, 196
disequilibrium, 175
disorder, 8
dispersion, 146
distribution, 25, 42, 74, 86, 132
divergence, 138
diversity, 155
DNA, 1, 2, 3, 4, 5, 6, 7, 8, 9, 10, 11, 12, 13, 14, 15, 16, 17, 18, 19, 20, 21, 22, 23, 24, 25, 26, 27, 28, 29, 30, 31, 32, 33, 34, 36, 37, 38, 39, 40, 41, 42, 43, 44, 45, 46, 47, 48, 49, 50, 51, 52, 53, 54, 55, 57, 58, 59, 60, 62, 63, 64, 65, 66, 67, 68, 69, 70, 71, 72, 73, 74, 75, 76, 77, 78, 79, 80, 81, 82, 83, 84, 85, 86, 87, 88, 89, 90, 91, 94, 95, 96, 97, 99, 105, 106, 112, 113, 114, 115, 116, 118, 120, 123, 124, 125, 127, 128, 129, 130, 131, 133, 135, 136, 137, 139, 140, 141, 143,145, 146, 148, 149, 151, 152, 154, 155, 157, 160, 162, 164, 165, 167, 170, 172, 173, 175, 177, 178, 179, 180, 181, 184, 186, 187, 191, 192, 193, 194, 195, 196
DNA damage, 18, 27, 49, 58, 89, 90, 118, 127, 146, 165, 181
DNA methylation, 1, 2, 3, 4, 5, 6, 7, 8, 9, 10, 11, 12, 13, 14, 15, 16, 17, 18, 19, 20, 21, 22, 23, 25, 26, 27, 28, 30, 32, 33, 34, 36, 38, 39, 41, 42, 43, 46, 49, 51, 55, 57, 58, 59, 62, 64, 65, 66, 67, 68, 70, 71, 73, 74, 80, 81, 83, 84, 85, 87, 88, 89, 90, 91, 94, 95, 96, 99, 105, 106, 112, 113, 114, 116, 119, 121, 123, 124, 125, 127, 128, 129, 130, 131, 133, 135, 136, 137, 140, 143, 145, 147, 149, 151, 152, 154, 155, 160, 164, 165, 170, 172, 173, 175, 177, 183, 184, 186, 191, 192, 193, 194, 195, 196
DNA methyltransferase enzymes (DNMTs), 30, 34, 37, 40, 46, 59
DNA methyltransferases, 1, 3, 24, 26, 28, 34, 68, 127, 136
DNA packaging, 47, 60, 62
DNA polymerase, 70, 184
DNA repair, 24, 29, 34, 37, 55, 58
DNA strand breaks, 83, 174

DNA-protein interaction, 36, 43, 46, 49, 54, 60, 114
DNase, 46, 77
DOHaD, 2, 142
donors, 13, 20, 28
dosage, 108, 132, 171, 172
dosage compensation, 108
doses of BPA induce, 4
double helix, 42, 75
double-strand helix, 42, 43
Down syndrome, 143
drinking water, 6, 8
drosophila, 58, 59, 87, 88, 108, 126, 167
drought, 163, 172
drug discovery, 159
drug targets, 117
dynamism, viii, 29

E

E26 transformation-specific (ETS), 30, 54
economic status, 59
ectoderm, 97
editors, 113, 124, 128, 139, 162, 164
EEG, 18
egg, 1, 3
electric charge, 53
electron, 147
electron microscopy, 147
electrophoresis, 116, 146, 183
elongation, 100
elucidation, 155, 171
embryogenesis, 24, 68, 93, 169, 191
embryonic stem cells, 15, 25, 28, 67, 68, 76, 78
encapsulation, 169
encoding, 17, 55, 155
endocrine, 2, 4, 5, 17, 19, 23, 24
endocrine-disrupting chemicals, 4, 28
endonuclease, 30, 46
endurance, 165
energy, 45, 121, 143, 164
energy expenditure, 143
energy transfer, 121
engineering, 165

environment, 6, 10, 97, 131, 134, 137, 138, 142, 155, 187
environmental change, 89
environmental conditions, 57, 62
environmental factors, 2, 3, 15, 16, 59, 130, 135
environmental stimuli, 2, 3
environments, 9
enzyme, 30, 32, 34, 37, 40, 41, 46, 50, 58, 59, 119, 127, 193
epidemiology, 2, 27, 129, 143
epidermis, 147
epigenetic alterations, 127
epigenetic modification, 3, 93, 96, 97, 130
epigenetic silencing, 133
epigenetics, 32, 34, 62, 72, 97, 105, 112, 113, 114, 127, 130, 132, 134, 136, 139, 140, 152, 162, 170, 171, 172, 173, 187
epithelia, 160
epithelial ovarian cancer, 66
Escherichia coli, 59
EST, 116
estrogen, 4, 28, 65, 165
ethanol, 14
ethers, 47
ethnicity, 59, 87
etiology, 140
euchromatin, 78
eukaryotic, 84, 147
evidence, 13, 16, 17, 21, 25, 33, 56, 93, 130, 139, 193
evolution, 73, 138, 141, 143, 144
examinations, 9
excision, 37, 58, 69
exercise, 59, 87, 165
exons, 42, 55, 56, 74
experimental design, 114
exposure, 2, 3, 4, 5, 6, 7, 8, 9, 10, 11, 12, 14, 15, 16, 17, 18, 19, 20, 21, 22, 23, 24, 25, 26, 27, 28, 80, 188, 192
expressed sequence tag, 116
extraction, 144, 160, 182, 186
extracts, 57
EZH2, 96, 97

F

fat, 13, 19, 26
fear, 13, 20
fertility, 17, 183
fertilization, 71, 181
fetal alcohol spectrum disorders, 14, 20
fetal development, 2, 92
fetal growth, 11, 142
fetus, 3, 13, 18
fibers, 121, 146
fibroblasts, 39, 40, 57, 65, 80, 81, 90, 97
fidelity, 67
Fiji, 146
fixation, 145, 159
flame, 18
fluorescence, 99, 121, 146, 159, 163, 181
folate, 13, 16, 17, 20, 165, 193, 194
folic acid, 13, 34, 68
food, 111
force, 46, 147
forebrain, 68
formation, 40, 43, 47, 48, 53, 67, 79, 85, 92, 96, 133
France, 123, 176
frontal cortex, 9, 41, 57
functional groups, 34, 45, 47, 55, 62
fusion, 109

G

gametogenesis, 38, 93, 94, 132
gel, 116, 146, 183
gender differences, 24
gene bodies, 33, 42, 44, 56, 84
gene expression, 6, 8, 10, 14, 16, 19, 23, 24, 32, 33, 38, 41, 47, 53, 56, 59, 66, 67, 70, 71, 84, 103, 112, 113, 116, 123, 125, 132, 134, 136, 138, 139, 143, 145, 149, 154, 155, 156, 159, 175, 185, 188, 193, 195
gene inactivation, 38, 40, 49, 185
gene promoter, 11, 12, 19, 33, 38, 42, 52, 57, 67, 94, 95, 96, 196
gene regulation, 1, 2, 30, 52, 57, 59, 62, 106, 107, 127, 132, 133, 136, 191
gene silencing, 29, 33, 67, 68, 95, 96, 128, 172, 174, 184, 192
gene targeting, 126
gene therapy, 92, 100, 108
gene transfer, 108, 183
genes, 3, 4, 6, 7, 8, 10, 11, 12, 13, 14, 15, 18, 22, 23, 25, 26, 32, 33, 38, 39, 41, 42, 44, 49, 53, 55, 56, 57, 59, 65, 66, 72, 80, 82, 85, 89, 92, 93, 94, 95, 102, 103, 107, 108, 116, 118, 127, 136, 137, 142, 150, 152, 163, 168, 172, 185, 195, 196
genetic code, 43
genetic disease, 95
genetic factors, 134
genetic information, 38, 62
genetic mutations, 3
genetics, 62, 112, 113, 115, 116, 124, 128, 129, 130, 135, 137, 140, 141, 143, 148, 150, 172, 173, 174, 175, 177, 179, 180, 184, 185, 186
genome, 8, 9, 15, 32, 33, 35, 36, 38, 40, 41, 49, 51, 52, 55, 57, 59, 62, 70, 71, 73, 80, 83, 84, 89, 112, 113, 114, 121, 126, 132, 136, 143, 145, 149, 152, 155, 165, 171, 172, 173, 175, 180, 182, 186, 193
genomic imprinting, 2, 3, 4, 18, 26, 32, 38, 56, 92, 93, 94, 102, 103, 104, 105, 106, 107, 151, 152, 171, 172
genomic instability, 16
genomic regions, 56, 62
genomic stability, 49, 193
genomics, 111, 112, 113, 117, 135, 148, 174, 175, 184, 185, 186
genotype, 116, 194
genotyping, 118
geometry, 45, 146
germ cells, 1, 3, 6, 40, 152, 153
germ line, 3, 4, 16, 21
Germany, 176
gestation, 7, 13
gestational age, 13
gestational diabetes, 12
gestational diabetes mellitus, 12, 25

gland, 5
glial cells, 56
glioblastoma, 104
glucocorticoid, 6, 18, 55, 84
glucocorticoid receptor, 6, 18, 55, 84
glue, 96
glutamic acid, 25
glutathione, 23, 123
gonads, 153
graduate students, 131
growth, 11, 13, 14, 18, 27, 31, 64, 66, 81, 91, 92, 93, 94, 100, 101, 102, 103, 104, 105, 106, 107, 109, 141, 142, 154, 163, 165
growth factor, 31, 91, 92, 93, 101, 102, 103, 104, 105, 106, 107, 109
growth hormone, 13
guanine, 15, 30, 44, 45, 46, 48, 56

H

hair, 146
hair follicle, 146
HBV, 104
HDAC, 83, 123
health, 2, 9, 16, 17, 19, 20, 24, 130, 134, 135, 142, 162
health effects, 2, 9, 24
heart disease, 142
heavy metals, 6, 9, 163, 170
HeLa cells, 57, 118
hepatitis, 66, 104, 162
hepatocarcinogenesis, 105
hepatocellular carcinoma, 27, 66, 103, 105, 165
hepatoma, 93
heritability, 143
heterochromatin, 3, 47, 49, 63, 78, 79, 133, 149
hippocampus, 7, 14, 26, 58
histidine, 46
histone, 7, 31, 35, 36, 44, 51, 53, 60, 62, 81, 82, 83, 85, 91, 96, 97, 98, 99, 105, 107, 108, 120, 127, 128, 131, 136, 147, 170, 172, 174, 177, 188, 195

histone acetylation, 7, 35, 36, 44, 82, 85, 131, 137
histone deacetylase, 83, 129, 195
histone H3 acetyl K14 (H3K14ac), 31, 76
histone H3 lysine 14 (H3k14), 31, 44, 76
histone H3 tri methyl K36 (H3K36me3), 31, 44, 74
histone H4 mono methyl K20 (H4K20me1), 31, 53
histone methylation, 44, 170
histones, 38, 42, 47, 51, 52, 62, 115, 121, 129, 132, 183
history, 77, 141, 142, 180
HIV, 136
HIV-1, 136
HLA, 31, 55, 84, 178
homocysteine, 20, 28
hormone, 4, 13, 21, 52, 165
host, 98, 100
hot spots, 6, 23
House, 56
housekeeping, 41, 54, 56, 57
HPV, 117, 118
human, 2, 6, 7, 9, 10, 15, 18, 20, 21, 24, 28, 31, 40, 41, 43, 46, 48, 56, 58, 59, 62, 63, 64, 65, 66, 68, 70, 71, 73, 74, 75, 78, 79, 80, 83, 84, 85, 86, 87, 92, 93, 94, 95, 96, 97, 98, 102, 103, 105, 107, 108, 109, 114, 115, 118, 120, 126, 129, 134, 136,139, 141, 143, 144, 147, 151, 153, 157, 165, 182, 187, 192
human genome, 62, 73, 84, 92, 136
human health, 134
human leukocyte antigen G (HLA-G), 31, 55, 84
husbandry, 186
hybrid, 102, 125
hybridization, 144, 154, 158, 159, 160, 174, 176
hydrocarbons, 4
hydrogen, 44, 45, 46, 47, 48, 49, 146
hydrogen bond, 44, 45, 46, 47, 48, 49
hydrophilic (polar), 47, 48, 52
hydrophobic (non-polar), 47, 48, 50, 53, 79
hydroxyl, 34, 45

hypermethylation, 6, 7, 8, 28, 39, 57, 65, 94, 96, 137, 140, 193, 195
hypertrophy, 8
hypomethylation, 5, 6, 7, 8, 13, 16, 18, 27, 28, 38, 51, 65, 66, 67, 82, 85, 86, 96, 104, 127, 192, 194, 195
hypothalamus, 14, 19
hypothesis, 2, 27, 29, 58, 172, 194

I

ideal, 48
identical twins, 32, 57
identification, 67, 116, 149, 151, 160, 182
identity, 68
idiopathic, 140
ILAR, 102
image, 121, 146, 147
immune response, 8
immune system, 65
immunodeficiency, 3
immunofluorescence, 89, 144
immunohistochemistry, 145, 154
immunoprecipitation, 147, 152, 154, 174, 186
imprinted maternally expressed noncoding transcript (H19), 10, 31, 38, 64, 71, 92, 93, 95, 98, 99, 102, 103, 104, 105, 106, 107
imprinting, 2, 3, 4, 13, 17, 18, 25, 26, 32, 38, 52, 56, 91, 92, 93, 94, 95, 96, 97, 98, 99, 100, 101, 102, 103, 104, 105, 106, 107, 108, 109, 151, 152, 171, 172, 177
impulsivity, 151
in situ hybridization, 123, 146, 158, 181
in utero, 13, 22, 23
in vitro, 53, 94, 118, 121, 123, 155, 169, 184
in vivo, 23, 49, 66, 83, 94, 120, 146, 152, 186
incidence, 92
India, 66, 113, 162
individual differences, 32, 57
individuals, 58, 65, 94
induction, 26, 81, 82, 100, 118, 123, 170

industrial emissions, 6, 8
infants, 13, 14, 22
infection, 99, 100, 119, 148
inflammation, 8, 22, 130, 137, 165
infrastructure, 114
inheritance, 26, 34, 51, 62, 81, 131, 134
inhibition, 57, 100, 123, 140
inhibitor, 8, 30
injury, 104
INS, 27
insertion, 155
insulin, 13, 19, 31, 39, 93, 94, 102, 103, 104, 106, 107, 109, 165
insulin resistance, 165
insulin sensitivity, 13
insulin-like growth factor 2 (IGF-2), 31, 91, 92, 101, 102, 103, 109
integrin, 118
integrity, 70, 181, 193
interface, 48
inter-genic regions, 42
inter-individual differences, 32, 57
interneurons, 25
intra-genic regions, 42
intra-individual differences, 57
intron, 74
introns, 42, 55, 56, 74
ion-exchange, 79
ions, 49
Ireland, 171
iron, 24
islands, 1, 2, 6, 9, 11, 12, 14, 18, 41, 56, 64, 65, 74, 80, 84, 95, 137, 195
isolation, 145, 147, 153, 182
issues, 59, 115, 124, 140, 165
Italy, 193

J

Japan, 1, 16, 166, 195
JUND, 31, 54, 55

K

kaiso, 48, 49, 50, 51, 75, 80
kaiso-like proteins, 49
kaiso-like zinc finger proteins, 48
keratinocyte, 39
keratinocytes, 118
ketones, 47
kidney, 39, 104
kill, 100
kinetics, 45
kirsten rat sarcoma viral oncogene homolog (KRAS), 31, 55, 84, 177
Korea, 164, 166
krüppel-like factor 4 (KLF4), 31, 49, 81

L

labeling, 144, 146
landscape, 81, 84, 136, 153, 171
larvae, 158
lead, 7, 8, 16, 21, 26, 93, 143
learning, 175
leptin, 13
lesions, 70, 119
leukemia, 89, 93
leukocytes, 8, 94
life cycle, 118, 135
lifetime, 32
light, 133
LINE-1 (long interspersed nuclear element 1), 10, 13, 31, 59, 65, 86, 192
liver, 6, 8, 39, 40, 57, 72, 92, 104, 161, 165
liver disease, 40, 72, 161
localization, 183
loci, 15, 38, 92, 95, 106, 147, 149, 152
locomotor, 9, 26, 27
locus, 11, 71, 96, 102, 104, 107, 172, 173
longevity, 140
lung cancer, 58, 66, 84, 85, 93
Luo, 65, 66, 172
lycopene, 165
lying, 43
lymphocytes, 65
lymphoma, 85
lysine (Lys), 31, 51, 53, 81, 82, 97, 107, 108

M

machinery, 49, 68, 93, 94, 97, 98, 100, 130, 136, 163
majority, 3, 38, 55
malignancy, 103
malignant tumors, 91
mammalian brain, 74, 142
mammals, 2, 32, 34, 42, 57, 71, 73, 86, 105, 108, 133
management, 169, 186
mangroves, 163
manipulation, 126, 186
manufacturing, 21
mapping, 85, 114, 132, 138, 156, 174, 186
mass, 38, 63, 145, 147
mass spectrometry, 145, 147
materials, 24, 117, 171, 177
maternal diet, 2, 12
maternal smoking, 23
matrix, 150
matter, 2, 3, 9
measurement, 89, 149, 163, 169, 193
mechanical properties, 76
MeCP2 (methyl CpG binding protein 2), 14, 15, 19, 21, 27, 31, 48, 49, 50, 52, 67, 79, 82
mediation, 187
medicine, 20, 128, 135, 142, 160, 162
mellitus, 12, 25, 92
melting, 45, 132, 172
melting temperature, 45
memory, 131
mental disorder, 140
mental health, 131
mental retardation, 7
mental state, 88
mercury, 7
mesenchymal stem cells, 127
messengers, 161
meta-analysis, 24
metabolic, 136, 165

Index

metabolic syndrome, 24
metabolism, 2, 8, 13, 39, 92, 115, 129, 165, 193
metabolites, 13, 21, 34, 47, 58, 102, 149
metal ion, 126
metals, 2, 3, 7, 9, 170, 188
metaphase, 147
metastasis, 85, 86, 161
methodology, 54
methyl, 7, 8, 13, 14, 15, 17, 20, 21, 26, 27, 28, 31, 32, 33, 35, 40, 43, 45, 47, 48, 49, 52, 53, 55, 58, 59, 68, 78, 79, 82, 88, 147, 193, 194, 196
methyl binding domain (MBD), 27, 31, 47, 48, 49, 50, 51, 52, 79, 80, 82
methyl binding proteins (MBPs), 31, 47, 48, 59
methyl group, 8, 13, 32, 33, 35, 45, 47, 48, 53, 55, 59, 193, 194, 196
methylation, 1, 2, 3, 4, 5, 6, 7, 8, 9, 10, 11, 12, 13, 14, 15, 16, 17, 18, 19, 20, 21, 22, 23, 24, 25, 26, 27, 28, 30, 32, 33, 34, 36, 37, 38, 39, 40, 41, 42, 43, 45, 46, 47, 48, 49, 51, 52, 54, 55, 56, 57, 58, 59, 62, 64, 65, 66, 67, 68, 70, 71, 72, 73, 74, 75, 76, 77, 80, 81, 83, 84, 85, 87, 88, 89, 90, 91, 94, 95, 96, 97, 98, 99, 104, 105, 106, 107, 108, 112, 113, 114, 116, 119, 121, 123, 124, 125, 127, 128, 129, 130, 131, 133, 135, 136, 137, 140, 143, 145, 147, 149, 151, 152, 154, 155, 160, 164, 165, 170, 172, 173, 175, 177, 183, 184, 186, 191, 192, 193, 194, 195, 196
methylation-insensitive TFs, 54, 55
methylation-sensitive TFs, 54, 55
methylation-specific TFs, 54, 55
Methylmercury, 7
methylome, 30, 34, 74, 83
mice, 6, 7, 8, 9, 13, 19, 20, 26, 27, 28, 38, 58, 69, 93, 94, 104, 105, 106, 119, 125, 152, 153, 157
microbial community, 155
microRNA, 14, 84, 114, 145, 149, 159, 160, 161
microscopy, 144, 146

mitochondria, 40, 72
mitochondrial DNA (mtDNA), 31, 37, 40, 41, 42, 70, 72
mitogen, 91
model system, 121
models, 9, 35, 94, 95, 100, 122, 131, 143, 150, 165, 174, 175, 176, 183
modifications, 29, 32, 33, 34, 35, 36, 43, 44, 45, 52, 53, 57, 59, 60, 61, 62, 89, 91, 95, 106, 120, 127, 136, 138, 139, 147, 166, 172, 174, 177, 188
modules, 126, 184
mole, 45
molecular biology, 32, 117, 120, 122, 124, 127, 145, 146, 148, 153, 154, 156, 158, 160, 168, 170, 173, 174, 177, 179, 183
molecular structure, 42
molecules, 16, 40, 46, 58, 59, 70, 184
monozygotic twins, 64
morphology, 43, 119, 181
mortality, 25
mother cell, 34
motif, 46, 53, 83
motivation, 150
mouse, 2, 4, 5, 6, 7, 9, 10, 14, 20, 21, 22, 23, 26, 34, 40, 56, 57, 63, 67, 68, 69, 71, 72, 74, 76, 78, 80, 81, 84, 86, 90, 93, 95, 96, 100, 103, 106, 119, 123, 146, 153, 157, 159, 165, 167, 183
MRI, 157
mRNA, 4, 6, 8, 14, 30, 85, 93, 158, 159, 182
mtDNA, 31, 37, 40, 41, 42
mucosa, 7
multiple factors, 30, 61, 62
multiple sclerosis, 161, 192
multivariate analysis, 175
Mus (musculus), 58
mutagenesis, 185
mutant, 120, 149, 194
mutation, 10, 16, 17, 18, 22, 45, 65, 91, 106, 139, 149, 173, 185
mutation rate, 16, 18
myelocytomatosis oncogene (Myc), 31

N

N6-methyladenine (6mA), 30, 59, 88
Na+, 46
NAD, 100, 112
nanomaterials, 10
nanoparticles, 10, 15, 18, 28
natural compound, 135
nervous system, 7, 33, 41, 142
Netherlands, 185
neuroblastoma, 81
neurodegeneration, 135, 140
neurodegenerative diseases, 38
neurodegenerative disorders, 135
neurodevelopmental disorders, 130
neurogenesis, 10, 142, 165
neuroinflammation, 9, 18
neurons, 22, 55, 56, 57, 68
neuroscience, 27, 151
neurotransmitter, 9
neutrophils, 24
next generation, 38, 154, 156
nicotine, 14
nitrogen, 35, 149
nodules, 27
non-CpG, 41, 56, 72, 74
normal aging, 8
normal development, 2, 3, 34, 56
Northern blot, 174
NRF, 54
Nrf2, 123, 166
nuclear DNA, 22, 37, 38, 39, 40, 191
nuclear receptor subfamily 3 group C member 1 (Nr3c1), 5, 31, 55
nuclear-respiratory factor 1 (NRF1), 31
nuclei, 86, 97, 147, 172, 183
nucleic acid, 160, 183, 184
nucleolus, 147
nucleoprotein, 35, 36
nucleosome, 41, 42, 43, 44, 52, 73, 74, 77, 120, 121, 132
nucleotide sequence, 3
nucleotides, 184
nucleus, 24, 38, 43, 47, 52, 97
null, 93

nutrients, 193
nutrition, 2, 17, 123, 134, 163, 164

O

obesity, 13, 130, 164, 192
obstruction, 15
oilseed, 112
oligomers, 43
oncogene, 57, 66, 81, 82, 85, 103, 104, 107
oncogenes, 33, 66, 105, 118
oncogenesis, 127, 195
oocyte, 5, 6
organ, 3, 183
organelle, 40
organism, 38, 40, 57
ornamental plants, 169
osmotic stress, 163
osteoarthritis, 128
ovarian cancer, 57, 84, 102, 165
ovarian tumor, 140
ovaries, 154
overweight, 24
oxidation, 26
oxidative damage, 15, 58
oxidative stress, 23, 27, 182

P

p53, 31, 33, 49, 66, 80, 102, 118
pachytene, 153
Pacific, 83
pairing, 45
palindromic, 53
pancreas, 39
pancreatic cancer, 85, 102, 116
pancreatitis, 165
pap smear, 119
parallel, 28, 132, 155
parallelism, 58
parenting, 92
participants, 10
particulate matter, 2, 3, 9
passive demethylation, 34

Index

pathogenesis, 92, 101, 117, 139, 149
pathogens, 41, 169
pathology, 117
pathophysiological, 25
pathway, 2, 13, 16, 23, 25, 34, 62, 73, 79, 92, 94, 98, 102, 116, 133, 175, 193
PBMC, 193
PCR, 28, 93, 100, 118, 145, 147, 155, 172, 174, 177, 181
PDX1 (pancreatic-duodenal homeobox factor 1), 31, 39
penis, 117
peptide, 93, 184
perinatal, 18, 22, 24, 154
peripheral blood, 65, 87, 94, 101, 193
peripheral blood mononuclear cell, 193
peroxisome proliferator-activated receptors (PPAR), 31, 52, 82
pests, 169
phage, 125, 185
pharmaceutical, 165
phenotype, 3, 65, 94, 116, 175
phenylalanine, 46
phosphate, 30, 32, 44, 46
phosphorylation, 138, 140
photobleaching, 121
photosynthesis, 41
phthalates, 4
physical activity, 164, 192
physical characteristics, 42
physical properties, 45, 47
physics, 30, 33
physiology, 141, 162, 163, 166, 184
phytoremediation, 169
PI3K, 92, 102
PI3K/AKT, 102
pi–pi (π–π) stacking interaction, 46
placenta, 6, 7, 10, 25, 56, 142
plants, 41, 112, 121, 131, 133, 152, 162, 163, 165, 169, 171, 172, 173, 192
plasmid, 147
plasmid DNA, 147
plasticity, 24, 137, 142
plasticizer, 23
plastid, 37, 40, 41, 73

platform, 145, 155
playing, 33, 51, 59, 139
pleiotropy, 142
PNA, 160, 184
point mutation, 149
polar, 47
polarity, 47, 52, 53
pollution, 9, 18, 23, 25
polychlorinated biphenyl, 18
polycomb group proteins, 108
polycomb repressive complex, 96, 97, 99, 105, 108
polymerase, 83, 116, 136
polymerase chain reaction, 116
polymorphism, 55, 84, 116, 134, 172, 174, 186, 193, 194
polypeptide, 30, 69
population, 8, 10, 84, 143
portraits, 115
positive correlation, 44, 56
postnatal exposure, 8
pregnancy, 2, 8, 12, 13, 18, 19, 21, 23, 24, 25
prenatal exposure to, 2, 3, 4, 6, 7, 8, 9, 10, 15, 16, 22, 26
preparation, 16, 148, 156, 182
preservation, 119
president, 180
prevention, 17, 117, 166
principles, 28, 124, 141, 143, 192
probe, 119, 146, 159, 186
professionals, 131
prognosis, 66, 91, 195, 196
programming, 13, 24, 27, 140, 143, 184
proliferating cell nuclear antigen (PCNA), 31, 51, 82
proliferation, 33, 91, 92, 103, 118
promoter, 8, 9, 11, 12, 14, 22, 25, 33, 41, 42, 44, 49, 52, 55, 56, 57, 59, 64, 65, 66, 68, 71, 73, 74, 76, 80, 83, 84, 85, 91, 93, 95, 96, 97, 98, 99, 100, 103, 105, 108, 140, 195, 196
propagation, 169
prophase, 154
prostate cancer, 85, 102, 116

protection, 28, 41
protein analysis, 150
protein family, 31
protein structure, 49
protein synthesis, 41, 73, 100
protein-protein interactions, 121
proteins, 26, 30, 31, 38, 39, 40, 42, 45, 47, 48, 49, 50, 51, 52, 53, 55, 59, 62, 63, 75, 78, 79, 88, 97, 100, 101, 107, 108, 112, 119, 121, 124, 125, 126, 188
proteome, 62, 182
proteomics, 116, 148, 164
proto-oncogene, 33, 57
psychiatric disorder, 130
psychiatry, 129
psychology, 151
psychosis, 161
PTEN, 94
public health, 28
purification, 120, 152, 153
pyrimidine, 33

Q

quality improvement, 112
quantification, 119, 146, 174
Queensland, 196

R

radiation, 67, 92, 100, 119, 188
Raman spectroscopy, 163
rape, 112
rat sarcoma virus (RAS), 31, 33
reactant, 58
reactions, 29, 34, 46, 57
reactive oxygen species, 15, 24, 123
reagents, 117, 171, 177
receptor, 4, 6, 7, 13, 28, 31, 52, 65, 83, 92, 102, 160, 164
recognition, 47, 48, 59, 74, 75, 78, 80, 81, 106, 184
recombination, 120, 174, 185
recovery, 121, 154

regeneration, 169
regression, 175
regulations, 40, 57
regulatory system, 93
relevance, 18, 66, 140, 160
remodelling, 146
repair, 18, 34, 37, 58, 69, 70, 127, 146
repetitive behavior, 27
repetitive DNA elements, 57
replication, 15, 22, 34, 35, 36, 38, 68, 99, 100, 119, 136, 147, 184
repression, 8, 28, 33, 49, 52, 53, 59, 79, 82
reproduction, 4, 28, 163
requirement, 62, 82
researchers, 161, 195
residue, 1, 2, 46, 48, 53, 53, 195
resistance, 92, 102, 127, 163
resolution, 77, 83, 132, 160, 172
resources, 18, 132, 170
response, 21, 23, 41, 49, 58, 84, 109, 112, 116, 118, 123, 163, 196
restriction enzyme, 46, 59, 73, 77, 120, 196
restriction fragment length polymorphis, 174
restriction modification system, 59
retardation, 14, 93
rheumatoid arthritis, 128
ribose, 100
ribosomal RNA, 136, 160
rings, 46
risk, 9, 10, 21, 25, 84, 94, 101, 105, 116, 117, 127, 134, 165, 195
risk assessment, 195
risk factors, 25
RNA, 16, 24, 34, 37, 41, 69, 70, 73, 83, 92, 93, 101, 103, 108, 114, 116, 119, 126, 132, 136, 145, 148, 152, 153, 155, 159, 166, 167, 168, 172, 174, 177, 182, 184, 187
RNA processing, 69
RNAi, 167
root, 126, 169

Index

S

s-adenosylmethionine [SAM (AdoMet)], 13, 26, 31, 34, 66, 136
safety, 111, 134
scanning electron microscopy, 182
schizophrenia, 25, 56, 84, 140
sclerosis, 65
seafood, 7
secondary metabolism, 149
selectivity, 54, 82, 184
semen, 182
seminiferous tubules, 182
senescence, 142
sensing, 126
sensitivity, 89, 100, 119, 159, 195
sequencing, 84, 113, 114, 132, 147, 149, 152, 154, 155, 156, 173
serine, 31, 48
serum, 13
services, iv
set and ring finger-associated (SRA) domain proteins, 31, 48, 50, 51, 79, 81, 82
sex, 4, 6, 18, 24, 56, 143
sex chromosome, 56
sex steroid, 143
sexual reproduction, 172
shear, 45
shores, 55
side chain, 47
silver, 10
silver nanoparticles, 10, 28
single-strand selective monofunctional uracil DNA glycosylase 1 (SMUG1), 31, 34, 69, 70
siRNA, 118
skeletal muscle, 136
skin, 165
small nucleoriboprotein n (Snrpn), 5, 10, 31, 38, 71
smoking, 15, 24, 25, 59, 151
smoking cessation, 15
SNP, 116, 145, 149, 178, 186
social environment, 16, 143
sodium, 196

software, 44, 114, 146
somatic cell, 38, 57, 78, 154, 186
Spain, 192
species, 15, 24, 58, 123, 148, 152, 172
specificity protein 1 (SP1), 31, 54
spectroscopy, 121
sperm, 4, 38, 87, 181
sperm function, 181
spermatocyte, 153
spermatogenesis, 95, 182
spermatogonial stem cells, 183
Sprague-Dawley rats, 18
Spring, 26, 180
stability, 45, 46, 59, 73, 81, 174
state, 10, 11, 12, 17, 64, 67, 77, 96, 114, 147
states, 140
statistics, 169
stem cells, 81, 93, 131, 142, 152, 153, 156
storage, 8
stress, 16, 22, 27, 41, 55, 58, 59, 62, 69, 70, 88, 89, 112, 123, 140, 142, 162, 163, 164, 172, 192
stress response, 22, 27, 172
stressors, 8
structure, 3, 30, 38, 42, 43, 44, 45, 46, 58, 60, 62, 74, 75, 77, 79, 81, 108, 120, 130, 136, 181
structuring, 47
substance use, 150
substance use disorders, 150
substitution, 193
substrate, 46
Sun, 22, 64, 66, 72, 73, 85, 112, 116, 137, 139, 155
supplementation, 13, 20, 27
suppression, 94, 96
survival, 91, 94, 101, 102, 105, 140
susceptibility, 2, 102
suspensions, 153, 169
swelling, 181
syndrome, 3, 17, 26, 68, 92, 102, 129, 143
synthesis, 69, 136, 174, 184

T

T cell, 8, 13
T lymphocytes, 7
tamoxifen, 92
target, 15, 21, 23, 59, 75, 77, 79, 82, 92, 98, 102, 108, 118, 125, 137, 185
techniques, 145, 163, 171, 174, 196
technology, 112, 114, 117, 126, 130, 132, 145, 149, 152, 155, 159, 177, 186
temperature, 132, 163
temporal lobe, 65
temporal lobe epilepsy, 65
ten-eleven translocation enzymes (TETs), 32, 37
teratology, 122
testis, 6, 39, 57, 153, 182
tetrachlorodibenzo-p-dioxin, 4, 5
therapeutic agents, 116
therapeutic approaches, 35
therapeutic effect, 100
therapeutic use, 184
therapeutics, 114, 161
therapy, 98, 99, 100, 109, 115, 135, 166, 195
thymine, 31, 34, 46
thymine DNA glycosylase (TDG), 26, 31, 34, 37, 63, 69
tissue, 3, 12, 15, 23, 25, 32, 38, 40, 57, 64, 70, 71, 72, 74, 85, 86, 93, 104, 116, 123, 144, 147, 148, 159, 169, 182, 192
tissue-specific gene expression, 38
toxic effect, 3
toxicity, 8, 10, 21, 27, 122, 125
toxicology, 18, 122, 124, 189
toxin, 98, 99, 100, 108, 109
traits, 171, 185
transcription, 1, 3, 8, 13, 18, 26, 30, 31, 32, 33, 38, 39, 40, 43, 52, 53, 54, 55, 56, 57, 59, 67, 74, 80, 83, 100, 109, 120, 125, 136
transcription activator proteins, 39, 53
transcription end site (TES), 32, 56
transcription factor 3 | achaete-scute homolog 1 (Tcf3|Ascl1), 31, 54
transcription factor JunD (a member of Jun protein family), 31
transcription factors (TFs), 32, 38, 40, 43, 52, 53, 54, 55, 59, 80, 125
transcription start site (TSS), 32, 56
transcriptional silencing, 3, 40, 52
transcripts, 153, 159
transduction, 8, 142, 152
transfection, 100
transfer RNA, 34, 70
transformation, 30, 41, 127, 170, 186
transgene, 104
translation, 34
translocation, 32, 127
transmission, 182
transplantation, 183
transport, 147
transposases, 126
treatment, 97, 98, 109, 116, 129, 195, 196
tributyltin chloride, 5, 20
troubleshooting, 117, 171, 177
tryptophan, 46
tuberculosis, 179
tumor, 8, 65, 66, 80, 85, 91, 92, 93, 94, 95, 96, 97, 98, 99, 100, 101, 102, 103, 104, 105, 106, 107, 108, 116, 195
tumor development, 195
tumor growth, 94, 100
tumor metastasis, 66
tumor progression, 66
tumor protein 53 (p53), 31, 33, 49, 66, 80, 102, 118
tumorigenesis, 66, 82, 92, 94, 101, 104, 105, 127, 135, 195
tumour suppressor gene, 33, 49, 57
Turkey, 29, 89
twins, 57, 130
type 1 diabetes, 130
type 2 diabetes, 130, 192
tyrosine, 31, 32, 46, 48, 55
tyrosine aminotransferase gene (TAT), 31

U

ubiquitin, 32

Index

ubiquitin-like, containing PHD and RING finger domains, 1 (UHRF1), 32, 50, 51, 81, 82
umbilical cord, 8, 22, 25
United Kingdom, 191
untranslated region (UTR), 32, 55, 84
urea, 147
urine, 24
USA, 44, 63, 70, 71, 72, 73, 75, 76, 77, 80, 83, 88, 104, 106

V

valine, 32, 48, 193
variations, 57, 176
vector, 100
vinclozolin, 4, 5, 21
virus infection, 162
viruses, 109, 177
vitamin B1, 34
vitamin B12, 34

W

weight gain, 18
Western blot, 97, 100
white blood cells, 94
World Health Organization, 181

X

X chromosome, 32, 38, 63, 64, 79, 108
X chromosome inactivation, 32, 38, 79
X inactive specific transcript, 32
xenografts, 119
Xist, 32, 38, 71
X-related genes, 38

Y

yeast, 58, 120

Z

zebrafish, 34, 57, 67, 69, 86, 122, 126, 133, 158, 167, 184, 185, 186
zinc, 14, 22, 32, 48, 49, 75, 78, 124, 125, 126, 185
zinc finger and BTB domain containing 33 (ZBTB33), 32, 49, 80
zinc finger protein 2 (ZF2), 32, 52
zinc finger protein 3 (ZF3), 32, 52
zinc finger protein 57 homolog (Zfp57), 15, 22, 32, 50, 52, 81
zygote, 38, 71, 87